A Master Framework
for the CRM Center of Excellence

Introducing universal standards for customer relationship management CoEs

By Velu Palani and Charlie Havens

A Master Framework for the CRM Center of Excellence
Introducing universal standards for customer relationship management CoEs
By Velu Palani and Charlie Havens

First paperback edition December 2024

ISBN 979-8-218-52137-0 (trade paperback)

www.crmcoe.com/book-bonuses

Praise for the Book

"The most successful CRM implementations have a Center of Excellence (CoE). But it is not something you buy and install. It is a set of principles, strategies, and standards. How you establish a CoE varies based on the structure of your CRM implementation and your organization. Therefore, you need a framework rather than a list of prescriptive tasks.

That is why this book is so powerful. It gives a series of frameworks and principles so that you can develop and evolve a CoE that will work for your situation. It is essential reading for CIOs, Center of Excellence leaders, and platform owners."

Ian Gotts, CEO and CMO
ElementsCloud.com

"I'm delighted to finally see a thorough guide to setting up and running a CoE. An effective CRM is central to every business, but the reality for most companies is that CRMs evolve organically without centralized oversight, guidance, or support. The most effective large organizations I've seen have taken the important step of establishing a CoE to provide the needed support functions. This is your guide."

Andrew Davis, Chief Product Officer of AutoRABIT
Author of *Mastering Salesforce DevOps*

"As a consultant and a CTO, I have seen organizations embark on transformation programs that leverage enterprise software, only to be disillusioned. Pitfalls included software not providing the proper functionality, cost to maintain and evolve the solution much higher than anticipated, time to delivery much longer than planned, etc. With SaaS solutions (like Salesforce CRM), we saw faster delivery. Still, many of the same pitfalls persisted, with the addition of a new pitfall, technology sprawl, i.e., many different independent implementations of solutions across the enterprise.

There is hope if you embark on an enterprise CRM journey or have inherited a complex, sprawling CRM environment. Having lived through the stand-up of a successful CRM CoE, I wish this book existed when we started. The book codifies many of the techniques that we found successful. The frameworks, principles, and advice provide much-needed guidance for anyone creating or managing a CoE."

Jim Petrassi, Divisional Senior Vice President & Chief Technology Officer
Healthcare Services Corporation

"Creating a successful CRM CoE is no small feat, requiring more than just technical expertise. This book is a master class defining the principles, strategies, and standards necessary to establish a robust CoE. It offers a flexible framework adaptable to any organization's unique CRM structure, making it an indispensable resource for CIOs, CoE leaders, and CRM platform owners."

Kalanithi Venkatesan, Founding Partner
4See Advisory LLC

Acknowledgments

I am grateful for the collective effort that has brought this project to fruition. Many people's expertise, insights, and dedication have been invaluable. From the initial brainstorming sessions to the countless hours of writing, editing, and revising, many contributions have shaped every aspect of this book. Charlie and I especially wish to acknowledge and thank the following for their invaluable contributions to the book's ideas and presentation: Jayneel Patel, Ph.D., Ian Gotts, Stella Michael, Andy Roy, Raghavendra Paripati, Scott Beach, Sachin Lande, Shanthi P. Palani, Sanjay Gidwani, Satish Padiyar, Jim Petrassi, Andrew Davis, Mach Lakshmanan, Mukesh Prusty, Vicky Skuja, Kim Gandhi, Amy Sanders, Sovan Bin, Carol Brundage, Pavan Reddivari, Hitesh Pushpraj, Peter Villeroy, James Kosmides, Richard Clark, Nicky Bang-Madsen, Sid Bhadani, Dan Pattangi, Ben Bradley, Phil Miller, Bruce Woll, Amy Osmond Cook, Jyothsna Bitra, Markus Gidlund, Vernon Keenan, Olivier Rachon, Elizabeth Southerland and my colleagues at HCSC.

I am particularly grateful to my co-author, Charlie Havens, whose deep understanding of the subject matter and passion for excellence helped create a comprehensive and practical guide.

To our editor, Carol Reed, thank you for your meticulous attention to detail, suggestions, and support throughout the writing process. Your guidance has been invaluable in ensuring our ideas are presented concisely and engagingly. Our illustrator and designer, Elizabeth Moss, brought our ideas to life with beautiful and creative designs. Our technical reviewers' thoughtful insights and suggestions helped us improve the book's quality and accuracy.

Finally, I want to thank my family and friends for their unwavering support and encouragement throughout this journey. Your belief in me and this project has been a constant source of motivation.

Without the contributions of these many people, this book would not have been possible. I am deeply grateful for our collaborative partnerships. I look forward to working with you as we move into future phases of this project.

Sincerely,

Velu Palani

Foreword

Greetings! I'm Sanjay Gidwani, Chief Operations Officer of Copado, Inc. My journey in the Salesforce eco-system began in 2008 when I started as a developer at Model Metrics, which was acquired by Salesforce in 2011. During my tenure at Salesforce, I collaborated closely with large enterprise customers and consistently encountered a significant challenge: organizing teams and processes to support substantial Salesforce initiatives effectively. This recurring issue sparked a deep desire to find solutions, ultimately leading me to Copado.

In 2020, I met Velu. I immediately recognized a person who shared my passion for solving complex puzzles in the Salesforce space. As a partner of Velu's, I can personally attest to his expertise and unwavering commitment to solving the challenges faced by CRM programs. His insights and guidance were instrumental in the success of our implementation, particularly in establishing a robust customer relationship management Center of Excellence (CoE) within his company.

In this invaluable book, Velu and Charlie generously share their wealth of knowledge, offering a practical guidebook for enterprises seeking to maximize the value of their software investments by creating and managing a CRM CoE. Drawing from their extensive experience, they outline a proven process successfully implemented at scale, addressing the critical need for organizations to extract tangible value from their software initiatives. In an era where businesses constantly strive to optimize their software utilization, this book provides a clear road map for establishing a CRM CoE that drives efficiency, innovation, and continuous improvement.

As companies embark on their journey to create or refine a CRM CoE, remember that the path to success is rarely a straight line. There will be challenges and setbacks, but the rewards of a well-implemented CRM CoE strategy are immense. Keep this guide as your trusted companion. Let its insights and experiences illuminate your path, and don't hesitate to revisit its pages as you navigate the journey ahead. Remember, the journey is just as important as the destination. As you scale your use of CRM and build your CoE, be prepared to adapt, learn, and evolve with your business.

Thank you for allowing me to be a part of your journey. As you go down your path, keep referring to this book. I'm confident it will continue to be a valuable resource for you.

I wish you the very best in establishing your CRM CoE. May your efforts transform your business and empower you to build stronger, more meaningful customer relationships and success as you scale your use of CRM and help drive your business transformation.

Sanjay Gidwani, COO
Copado, Inc.

Preface by Velu Palani

My journey with Salesforce began on June 13, 2000. I purchased one of the Salesforce licenses and used it to run my Oracle consulting business. Since then, I have held various Salesforce-related positions, including as a technical expert and CEO of a Salesforce consulting organization. I have developed tailored customer relationship management solutions for clients as a consultant for small, medium, and enterprise-size organizations. I am pleased to have assisted in establishing several CRM Centers of Excellence (CoE). Though each CoE faces unique challenges, the fact that no universal CoE framework existed from which to address those challenges was a great puzzlement and frustration.

It is our aim, with this book, to offer a comprehensive guide to establishing a robust customer relationship in the management of a CoE. Most importantly, it envisions a centralized hub for all CRM-related activities, ensuring that everyone involved is aligned and collaborates effectively to deliver an exceptional customer experience.

Whether you're a CoE leader, a decision-maker, or a stakeholder, this book is a goldmine of valuable insights and practical strategies to help your team flourish. But the book's aspiration goes beyond its immediate aim—it seeks to establish a standardized approach to CRM CoEs, something that is currently lacking. By creating a common language and framework, we can foster improved communication and collaboration within the CRM field.

A CRM CoE is the team assigned to systematically drive business outcomes via the CRM platform(s). Effectiveness demands productive collaboration with external stakeholders. This book guides aligning various stakeholders and facilitating seamless teamwork.

I have established an online community known as CRM CoE (accessible at www.crmcoe.com). The primary objective of this initiative is to accumulate valuable insights from diverse individuals within the industry and to ensure that all participants remain informed of the latest advances in the field. We intend to provide regular updates aimed at assisting businesses in optimizing their CRM investments, thereby facilitating substantial growth. I welcome your participation.

Join us on this transformative journey to revolutionize CRM CoEs. Let's build a collaborative support system that empowers businesses to thrive.

Preface by Charlie Havens

My journey with Velu began when I was contracted to teach Oracle developers how to pass Salesforce .com certification exams. Velu taught me that the core purpose of Salesforce.com is to produce business outcomes. This kernel became the cornerstone of our collaboration and the central theme of this book—that CRM CoEs exist to drive business outcomes.

This book addresses the critical questions that drove our projects: How can CoEs drive efforts toward producing business outcomes? How do we measure progress? And how do we ensure our efforts are economically justifiable? These questions have been our compass, leading us from project to project, CRM to CRM, and now, to this book.

Our work has thrived in an environment of trust, allowing us to explore, experiment, and iterate our solutions. This iterative process has enabled us to develop a framework that fills a significant gap in the CRM ecosystem. We don't just invite you to join us in this conversation; we urge you to bring insights and experiences to help refine and perfect this framework. This book is a collaborative effort, and your contribution would be invaluable.

We invite you to engage with this framework. We encourage you to guide us with your expertise, adapt it to your industry, and contribute to the evolving knowledge around CRM CoEs. Together, let's elevate CoE outcomes.

Table of Contents

Table of Figures

Building a CRM Center of Excellence by Velu Palani

As a seasoned Salesforce professional with 24 years of experience, I've worn multiple hats, from building products on the Salesforce AppExchange to leading professional services organizations, serving as a strategy consultant and implementing large-scale solutions for clients. I have always made it a priority to deliver high-quality projects within budget, but one aspect that often went overlooked was how my customers would manage their Salesforce implementations once they were complete. I developed solutions for numerous customers based on my developed standards and framework, but I was never able to be present afterward to see how the implemented solution was managed.

In 2020, I faced a pivotal test moment. I joined an organization in a leadership role to establish, manage, and demonstrate the value of a Center of Excellence (CoE) for its extensive Salesforce ecosystem. Initially, I assumed my experience and expertise would make this task straightforward. I could simply replicate what I had done before. However, the reality was much more complex. Large organizations each operate with unique organizational structures, cultures, politics, and frequently changing leadership. This presented new challenges that my previous structure did not adequately address.

Establishing an efficient CoE operating model, forming a skilled team, and educating leaders on the significance of security, governance, scalability, and resilience proved exceptionally challenging. While consulting with various global systems integrators, I observed that their CoE frameworks and those I implemented for previous clients didn't fully account for how clients would manage their systems after implementation. This new awareness underscored the need for a framework that tailors the process for each client's unique needs, ensuring they are fully equipped to manage their systems effectively.

With this realization, I embarked on developing a comprehensive global customer relationship management (CRM) CoE framework to help companies maximize their CRM investments. My goal was to establish a flexible standard CoE framework to address the industry-wide trend of failing to capitalize on the full benefits of CRM implementations. The lack of uniform standards and accountability for managing the completed implementation perpetuated subpar results.

Driven by the pressing need to break the repetitive cycle of inefficient, unsustainable, custom CRM CoEs, I was inspired to write this book. It not only provides guidance for designing and implementing a CoE but also addresses essential aspects, such as team assembly, success measurement, and the development of strategies for supporting business growth while minimizing risks. With a focus on practical utility, I collaborated with Charlie Havens, an expert in this field, to enhance the book's comprehensiveness. Our aim was to equip companies with a robust system that would enable them to maximize their CRM investments and drive sustainable growth.

Introduction

This book is designed to guide executives and customer relationship management (CRM) Center of Excellence (CoE) leaders in maximizing the impact of their CRM initiatives. In today's business landscape, aligning CRM with an organization's core mission and strategic goals is vital for growth, customer experience enhancement, innovation fostering, and realizing value from CRM investments. CRM systems are pivotal in managing customer relationships by providing an effective structured customer relationship management framework.

This framework emphasizes the significance of CoEs in achieving improved business outcomes, long-term success, and maximized value from CRM initiatives. We lay out a holistic framework for understanding, creating, and evaluating CRM CoEs.

The book covers various aspects of the broader issue, including defining a CoE and its purpose, the CoE maturity process, cost management, value realization, governance, leadership evolution, and digital transformation strategies. The framework ensures strategic alignment and maximizes CRM value.

Even if you're a seasoned executive overseeing CRM strategy or an experienced leader managing a CoE, you will find insights, methods, and practical tools. We provide advanced strategies and best practices to enhance and refine existing CoEs. Some topics you might have already mastered and find to be too rudimentary; others may surprise you and provide new ways of looking at how and why you do what you do. Because your system is different and it has worked for you, you may disagree with some points of this framework. However, by adopting this framework, regardless of how mature yours is, your teams will have adopted a common cross-domain language and process for identifying areas needing improvement and discussing that with other CoE teams. In short, you will have a common cross-domain framework for continuous alignment with organizational goals that foster value realization. This book helps optimize the CoE's performance, drive further innovation, and achieve sustained business success.

What if your company hasn't yet established a CoE—will this book be helpful? Absolutely. This book is an invaluable resource for organizations at any stage of CoE development. Suppose you are in the initial stages of establishing a CoE. In that case, this framework will guide you through establishing your CoE's specific purposes, defining where to place the CoE within your organization, and how to identify, leverage, and influence functions within your CoE's realm of interest. It helps define leadership roles and responsibilities, track progress with metrics, and align CRM initiatives with business objectives. The detailed insights and practical advice will help you build a robust and effective CoE from the ground up, ensuring long-term success and maximizing the value of CRM investments.

How to read this book

Individuals playing different roles within an organization can use the book in various ways.

Practitioners, leaders, and strategists: This book provides a comprehensive guide for CRM CoEs, offering valuable information tailored to their role. It provides actionable insights, strategic guidance, and long-term goal planning.

Initiators: Individuals identifying the need for a CRM CoE will find guidance on proposing and establishing one.

Funders: Those securing financial resources will gain insights into optimizing CRM functions economically and understanding the detailed plans needed to justify and optimize investments in the CoE.

Sponsors: Senior executives championing the CRM CoE will discover methods to secure organizational buy-in and provide the visibility and authority needed for implementing change.

Evaluators: People responsible for assessing the CRM CoE's performance will find tools for monitoring outcomes, analyzing data, suggesting improvements—all aimed at ensuring the CoE aligns with strategic business objectives.

CoE leads: CRM CoE managers will benefit from detailed chapters on governance structures, the maturation process, and leadership evolution. These will help them align CRM initiatives with company goals and ensure team success.

Readers can approach the book in different ways.

As a reference: The book provides actionable insights and best practices for practitioners to refer to as needed, like a cookbook.

Cover-to-cover reading: Leaders and strategists may benefit from reading the book from cover to cover to fully grasp its comprehensive nature and strategic direction.

Executive summaries and glossary: For a quick overview, some may prefer reading the executive summaries for each chapter, utilizing the glossary, and scanning the images, tables, and stories for critical insights.

The authors invite feedback from the CoE community to iteratively improve the framework over time. It is presented as a flexible theory to be proven by its users. The framework can be adapted to specific contexts, such as financial or healthcare companies and nonprofit or government organizations.

Although the book's examples are derived from the Salesforce.com ecosystem, it aims to be CRM-agnostic. Whether you use Salesforce, Microsoft Dynamics 365, SAP CRM, Oracle CRM on Demand, or another CRM platform, you are encouraged to engage with the central proposition, argue with it, prove it, improve it, and adapt it to your platform(s).

What is a CRM Center of Excellence?

In the dynamic landscape of modern business, the CRM Center of Excellence (CoE) role has emerged as a beacon of strategic visionary leadership and operational excellence. Its primary mission is to maximize the value derived from the organization's CRM investments while minimizing risks and inefficiencies. A CRM CoE represents the nucleus of an organization's CRM strategy. The CoE embodies a commitment to leadership excellence. This centralized entity defines and upholds CRM implementation principles, best practices, standards, governance, performance metrics, and shared use of tools. The CRM CoE is more than just an idea; it is a team of skilled individuals operating within a well-structured framework to synchronize CoE's realm of interest to systematically drive business outcomes via the CRM platform(s).

The Six Pillars of development

The Six Pillars of the CRM CoE are the primary work of the Center of Excellence. In contrast to the CoE's realm of interest, which encompasses a broader range of functions beyond the control of the CRM CoE, the Six Pillars are the CoE's core responsibilities and accountabilities. To the extent that these six pillars are well executed, they are the main drivers of the CoE's effectiveness. Each actively contributes to the management and strategic alignment of CRM initiatives with the organization's core mission, ensuring that CRM practices support the company's strategic goals and facilitate the realization of value from CRM investments.

These are the Six Pillars of development:

1. **Drive strategic alignment.** CoEs align CRM deployments closely with the company's strategic goals. Integrating CRM initiatives with the core mission, the CoE delivers consistent customer experiences and optimizes staff efficiency. This alignment ensures that every customer interaction supports the organization's long-term vision and strategic objectives.

2. **Establish governance for best practices, strategies, road maps, architecture, and tools**. A CRM CoE establishes and monitors policies, standards, guidelines, KPIs, and architecture designs related to CRM

implementation, change management, data governance, and risk management strategies. This role is vital for mitigating risks associated with legal, regulatory, and security requirements, enhancing organizational resilience and safeguarding against potential threats and liabilities. Additionally, it ensures high data quality and effective data management practices, which are crucial for maintaining accurate and reliable customer information.

3. **Measure ROI realization**. By streamlining processes, fostering user adoption, and guiding organizations through change management, the CRM CoE ensures smooth transitions and secures buy-in from critical stakeholders. This role is crucial for realizing returns on investment by maximizing the efficiency and effectiveness of CRM initiatives, thereby driving financial and operational benefits.

4. **Incorporate collaboration and interoperability**. The CRM CoE breaks down silos within the organization, fostering cooperation among different departments and teams. This role ensures data and system interoperability, aligning CRM efforts with broader digital and business goals. Enhanced collaboration leads to more cohesive and unified customer relationship management strategies, improving overall organizational performance.

5. **Hone strategy stewardship and innovation**. The CoE proactively assesses and adjusts strategies and processes to keep CRM efforts fresh, relevant, and aligned with organizational goals. The CRM CoE enhances organizational agility and customer engagement effectiveness by fostering a culture of innovation and continuous improvement. This role ensures that CRM initiatives remain dynamic and responsive to changing market conditions and customer needs.

6. **Nourish skill acquisition and expertise sharing**. Serving as a repository of expertise, the CRM CoE addresses skill gaps within the organization by promoting the continuous development of CRM skills across teams. This role ensures that all stakeholders can handle current and future CRM challenges. The CoE enhances the organization's overall capability to manage its CRM initiatives effectively by fostering constant learning and expertise sharing.

These Six Pillars collectively contribute to the strategic, coordinated, and optimized execution of CRM initiatives. In today's business environment, where multi-instance CRM systems play a central role in customer engagement, sales, marketing, and service, effective coordinated stewardship of these systems through a CoE is essential for a company's sustained success in the digital age.

Generative AI and CoE leadership

We can't talk about the role of the CoE without putting it in the context of generative Artificial Intelligence. AI's presence is all-pervasive. Staff positions are being reduced because of it, code is being written by it, AI is being used to develop user stories, and customer service voices are being generated. The omnipresence of AI is occurring so fast that we aren't clear when and whether it is a good thing or harmful. Are we implementing it fast enough? Are we using it where it isn't appropriate or ready?

The digital revolution

The digital revolution represents a transformative series of phases fundamentally altering how society functions, communicates, and processes information. This sweeping change has been gradual, encompassing several decades marked by pivotal technological breakthroughs and cultural shifts.

1980s	1990s	2000s	2010s	2020s
Introduction of personal computers and mobile phones	Birth of the World Wide Web	Growth of digital technology enhances media consumption	Internet becomes widely accessible across the world	Technology evolves rapidly and AI changes content creation

Figure 1: The trajectory of the global digital revolution.

In the 1980s, the digital landscape shifted dramatically with the introduction of personal computers into households and businesses. This decade experienced the transition from analog to digital, fundamentally altering work and entertainment paradigms. The era also saw the advent of the first mobile phones, albeit in their most basic forms, hinting at future communication possibilities.

The 1990s ushered in the internet era, characterized by its commercialization and the birth of the World Wide Web. This period revolutionized information sharing and commerce, with email emerging as a new standard for communication, breaking down the traditional barriers of time and distance.

Digital technology became a global phenomenon during the 2000s. The internet and mobile phones became ubiquitous, and this decade witnessed the transition of television broadcasts from analog to digital, significantly enhancing media consumption quality.

By the 2010s and beyond, connectivity had reached unprecedented levels, with the internet accessible to over a quarter of the world's population. The proliferation of smartphones and the advent of tablets challenged the dominance of personal computers, signaling a shift towards more portable and user-friendly devices for accessing digital content.

The disrupting nature of generative AI

Generative AI has dramatically disrupted traditional industries by automating creative and analytical processes previously dominated by humans. This paradigm shift has revolutionized how content is created, decisions are made, and innovations are conceptualized, offering unprecedented efficiency and personalization. However, it also raises significant ethical, economic, and social questions, as its pervasive influence redefines job roles, intellectual property rights, and the very nature of creativity and human contributions in the digital age.

The disruptive role of AI in CRM application delivery and the future of end-users

Generative AI is on the brink of revolutionizing CRM application delivery by enabling a more integrated and efficient approach. A skilled architect with deep industry and business knowledge can oversee the entire process, from writing user stories to deploying code and configurations. This holistic approach can eliminate technical debt and reduce total cost of ownership. By leveraging AI, architects can ensure that all stakeholders are aligned, leading to more coherent and effective CRM solutions. AI tools can assist in writing and optimizing code, reducing the time and effort required for development and ensuring that best practices are consistently applied, and potential issues are identified early. Predictive analytics allow for proactive decision-making and resource allocation by analyzing historical data to predict future trends and user behavior. AI-driven testing tools can simulate a wide range of scenarios and detect issues that might be missed by human testers, leading to more robust and reliable applications. Furthermore, AI can continuously monitor the performance of CRM systems and suggest improvements, ensuring that the system evolves with changing business needs.

For end-users, such as sales and service professionals, generative AI is transforming productivity and customer engagement. Sales professionals traditionally spend most of their time on mundane activities rather than collaborating with customers to identify problems, while service professionals spend excessive time documenting conversations. AI addresses these issues by analyzing customer data to identify high-value customers and prioritize them for follow-up, ensuring that sales efforts are focused on the most promising leads. It also helps create personalized and contextually relevant emails and messages, improving engagement and conversion rates. Most important, AI can transcribe and document audio and chat conversations in real time, significantly reducing the administrative burden on service professionals and allowing them to focus on customer interactions. Moreover, AI can predict potential issues before they arise, enabling service teams to address them proactively and improve customer satisfaction. This potential for improved user experience is a reason for excitement about the future of CRM application delivery.

As we write this book in 2024, AI has already started to revolutionize every aspect of customer relationship management, including CRM delivery and end-user management. CoE (Center of Excellence) leadership, responsible for driving best practices and innovation within an organization, must plan around the capabilities of AI when designing and implementing their strategies. This involves utilizing AI to gather and analyze data across the organization, providing insights that inform strategic decisions and drive continuous improvement. Ensuring that all team members are equipped with the necessary skills to leverage AI tools effectively fosters a culture of innovation and continuous learning. The role of AI in driving continuous improvement is a reason for excitement about the future of CRM application delivery.

AI is not just an add-on to existing CRM practices but a transformative force that requires a fundamental shift in how organizations approach CRM application delivery and end-user engagement. By embracing AI, organizations can achieve greater efficiency, deeper insights, and a more personalized customer experience, ultimately driving growth and competitive advantage.

CRM alignment with digital transformation

Aligning customer relationship management with digital transformation is pivotal for businesses looking to modernize and personalize their customer interactions. This alignment enhances the customer experience through data-driven insights and automation and ensures a cohesive strategy across all customer touchpoints. Such integration fosters a more agile, responsive, and customer-centric business model, which is crucial for thriving in today's fast-paced digital landscape.

The evolving role of CRM CoE leadership

Leading a CRM Center of Excellence has become increasingly complex. It now requires navigating rapid technological changes, organizational dynamics, and shifting customer expectations. Today's leaders must understand technology and exhibit strong skills in change management, strategic visioning, and cross-functional collaboration. They are tasked with championing the integration of CRM systems within the broader digital transformation strategy, ensuring alignment with business goals and fostering a culture of continuous improvement and innovation.

Future opportunities and trends

While it is challenging to predict future trends due to the disruptive nature of generative AI, the age of Enhanced Relationship Automation (ERA) is already here. It happens around and to us in ways we do not fully appreciate. Within the CRM CoE realm of interest, this focus leverages automation and AI to deepen customer relationships and streamline processes, emphasizing predictive analytics, personalization at scale, and enhancing customer experiences through seamless integration across touchpoints. By adopting ERA technologies, CoEs can drive innovation, anticipate customer needs, optimize customer journey mapping, and enable proactive engagement strategies, thus creating a competitive advantage and encouraging loyalty in a dynamic market landscape.

Still, we need humans to determine whether AI's outputs are valid and to interact with them to keep them properly focused. We need humans to determine appropriate uses and identify those not yet ready for prime time. We need frameworks that allow humans to collaborate better and guide them regarding when to use or not use such tools and to what ends. The framework of this book, by defining what the CoE should be about, allows better conversations among human actors regarding their own uses of generative AI and the role of CoE leadership in its adoption.

By providing guidance, standardization, and expertise, the CoE ensures that CRM strategies are aligned with company objectives and industry best practices, and ultimately contribute to long-term growth and prosperity. Furthermore, as detailed in Chapter 4, CoEs can enhance their impact by providing top-line revenue justifications and measurements for funding strategic CRM initiatives to drive revenue growth and improve customer satisfaction. This economic focus reinforces the CoE's role as a critical player in an organization's financial strategy.

The purpose of a CRM CoE

A CRM CoE is the team assigned to systematically drive business outcomes via the CRM platform(s). A customer relationship management (CRM) Center of Excellence (CoE) improves business outcomes and long-term success by aligning IT solutions with the organization's core mission and strategic goals and ensuring business value from those initiatives. A CRM CoE enhances business outcomes and long-term resilience by aligning its solutions with the larger organization's fundamental mission and strategic targets. It also guarantees the commercial value of these endeavors. And it improves outcomes by harmonizing strategic visions and governance across all functional areas of interest. This includes areas controlled by the CoE, those impacting the CoE, and those influenced by its actions. By establishing a CoE, organizations can systematically improve their CRM capabilities, leading to tangible benefits, such as increased customer satisfaction, reduced operational costs, enhanced revenue growth, and a clear value realization from CRM investments.

The Tale of the Sprawling CRM Ecosystem
by Charlie Havens

Once, upon joining a new company, I was asked to join a peculiar meeting devoid of context or introduction. Various owners of CRM instances gathered around the table, with the meeting helmed by a recently appointed head of marketing. It quickly became evident that this was their first effort to resolve a problem that had not yet been articulated or generally understood. The goal of the head of marketing was straightforward. Still, the goal itself was proving ambitious: Unify disparate CRM instances so that marketing could send out cohesive messages to sets of targeted customers across those instances. The execution was anything but simple. Diverse CRM architectural and data structures with similarly labeled fields and dissimilar values created a Tower of Babel, with the various CRM business support and IT teams as varied as the CRM architectures. I observed, in silence, as a newcomer lost in the situation's complexity.

A casual conversation during a break revealed the root of the chaos: a history of acquisitions and divisional autonomy had led to a labyrinth of CRM systems. Some predated current leadership, and others resulted from departments bypassing IT for quicker solutions, leading to an organic sprawl of systems, each supported by different external consultants. The aim was now to consolidate these systems, a task to which I was assigned.

Those intimately involved seemed as lost as I felt. The head of marketing seemed stunned by what he was discovering. No one could explain why things were as they were, highlighting the urgent need for a comprehensive review of the current CRM systems.

Different parts of the company were all doing their own thing, setting up their own customer relationship management systems without much oversight. Each had been led and the architecture designed by their CRM consultants, without cohesive central coordination across CRM systems.

This led to a crazy mix of CRM systems reporting to different company-siloed units, all designed with different architectures and incapable of talking to each other. It was like a zoo, with each system being a separate species of animal, each doing its own thing and unable to communicate with its neighbor. Marketing tried to get everyone on the same page, but it was trying to herd cats.

When we regrouped, we decided to align two of the CRM systems and then build upon that with the others, one at time. The initiative was laborious—creating a cohesive whole from components that had been designed from different perspectives.

This story, while extreme, typifies the challenges prompting the creation of CoEs in managing the CRM universe. Often, the diversity in CRM systems stems from legitimate, distinct needs, but problems arise when collaboration becomes necessary. The call for a CoE is a crucial solution to unify these disparate elements under a common goal, thereby streamlining CRM management and improving collaboration.

However, the absence of a standardized framework or language for defining the CoE's role, performance evaluation, funding, and authority leaves leaders to navigate these waters with little guidance. Success or failure hinges on the leader's ability to craft a vision and influence others to adopt it, often building on the foundations—or ruins—left by predecessors.

Looking outside their organization's experience for CoE frameworks and guidance, persistent CoE leaders find consulting firms also lack a universal standard. Each consulting company champions its bespoke approach without a collective dialogue on best practices. Here we are, without an industry-standard framework, despite spending billions of dollars on this CRM technology.

We as an industry have abdicated our responsibility to establish universal standards, leaving organizations to each come up with their duct-taped versions. And they do. Sometimes, there are many versions within the same company, each the abandoned stepchild of the different consulting companies and different initiators within the organization. Unlike the Agile domain, where frameworks like SAFe, Nexus, and others provide a common starting point and where DORA (DevOps Research and Assessment) offers metrics for benchmarking maturity, the CRM CoE realm lacks such a standardized toolbox.

This tale highlights the need for a common framework to navigate the chaos, a sentiment that inspired the creation of this guide. Stepping into that initial, bewildering meeting, I needed the toolkit and language this book aims to provide.

CHAPTER 1
A Framework for the CRM Center of Excellence

Executive summary

Chapter 1 introduces a universal framework for the Customer Relationship Management (CRM) Center of Excellence (CoE). This structured approach plays a role similar to that played by methodologies like ITIL[1] for IT service management or PMBOK[2] for project management. This framework integrates CRM Centers of Excellence (CoEs) with overarching business objectives, aligning customer relationship management with the organization's core mission and strategic goals. Doing so ensures that CRM initiatives drive significant business outcomes and long-term success, a focus that sets it apart from other framework.

The CRM CoE Framework is not just a theoretical concept but a practical tool that provides a unified language for assessment and enhancement. It encompasses principles, best practices, standards, and key performance indicators (KPIs), which can be directly applied to CRM initiatives. Adopting this framework ensures consistency, facilitates benchmarking, enhances collaboration, improves evaluation, guides best practices, supports training and development, accelerates implementation, and encourages growth. These practical benefits make the framework essential for constructing, maintaining, and elevating CRM CoEs, consistently delivering value within the larger business ecosystem. This chapter emphasizes driving value through CRM enhancements across functions for which you may be responsible or accountable or over which you may merely have informal influence.

One of the critical components of the CRM CoE Framework is the highly advantageous Five-Step Process for CRM CoE refinement. This strategic process addresses immediate operational needs while laying the foundation for long-term success and adaptability in a dynamic business environment. By following this structured progression, organizations can achieve critical business outcomes, such as growing revenue, increasing customer satisfaction, reducing operational expenses, improving user adoption, and lowering risk. The Five-Step Process is a crucial tool that can guide CRM initiatives toward success.

---

1   AXELOS, ITIL Foundation: ITIL 4 Edition (Norwich, England: The Stationery Office, 2019).
2   Project Management Institute, *A Guide to the Project Management Body of Knowledge* (PMBOK Guide), 6th ed. (Newtown Square, PA: Project Management Institute, 2017).

Defining a CRM CoE Framework

The CRM CoE Framework is a structured approach designed to optimize and effectively manage customer relationship management initiatives. The CRM CoE Framework is like other frameworks, such as ITIL, which offers best practices for IT service management and the Project Management Body of Knowledge (PMBOK), which outlines standards for project management. But the CRM CoE Framework differs in the following way: It integrates the operations of CoEs with overarching business objectives. It serves as a strategic tool, especially for enterprises with matrixed environments, furnishing a unified language for assessment and enhancement across diverse functions. This framework encompasses principles, best practices, standards, key performance indicators (KPIs), and a method for testing whether this framework and its measurements are addressing the right components.

Why have a universal framework?

A common industry-standard framework is pivotal for several reasons. First, it ensures consistency by establishing a shared language and understanding within and across companies, thereby reducing misalignment and confusion. Second, it facilitates benchmarking, allowing companies to measure their CRM practices against a recognized standard and encouraging continuous improvement. Additionally, a common framework enhances collaboration, aiding smoother cooperation and knowledge sharing among CRM CoEs based on shared principles and practices.

Adopting a universal structured framework is essential for understanding our specific reality. We have something external to compare to and as we make those comparisons across organizations using a common language, we will develop better ways of measuring and defining what makes for a high performance CoE. For example, by pinpointing the essential CoE functions that match specific stages of maturity across organizations, we can better guide the prioritization of specific efforts. Such information helps leaders make informed decisions and allocate resources efficiently to bridge gaps and address challenges within their organizations.

The framework is not just a guide but a key to unlocking the best practices for CRM CoEs by comparing what works in different environments, against a common structure. It assists us, as an industry, in adopting effective strategies, avoiding typical pitfalls, and fostering successful customer relationship management systems. It raises the industry standards for value realization by minimizing organizational risks, enhancing operational efficiency, boosting customer satisfaction, and driving overall revenue growth. This framework provides a source of confidence for constructing, maintaining, and elevating CoEs to consistently deliver value within the larger business ecosystem.

Control: Striking a balance

The operational scope and structural nuances of CRM CoEs vary significantly among organizations. Some manage a single CRM production instance, while others oversee a complex network of multiple production instances, each with numerous sandbox environments governed by different business segments or jointly by IT and business units. This variability in management and operational approaches highlights the diverse strategies that businesses deploy in CRM initiatives.

Balancing responsibility, accountability, and authority with the power of informal influence is critical in driving value through CRM enhancements. The CoE shouldn't seek direct control over all functions within its realm of interest. Still, the synergy and mutual influences of these functions are salient for superior business results. Recognizing and leveraging these interdependencies is crucial for effectively utilizing the CoE's influence.

Informal influence is the power to influence derived from personal attributes or relationships rather than formal authority. It recognizes that individuals or groups can significantly impact decisions and actions based on personal qualities, expertise, or relationships, even without formal leadership roles. Understanding these dynamics is essential for effectively navigating the CoE and broader company interactions.

We strategically deploy formal and informal influences to maximize the CoE's effectiveness in achieving business outcomes. We harness informal influence to drive changes and reach consensus more effectively, especially in matrixed organizations where formal authority is dispersed.

The CoE and its CRM environments

Before delving deeper into the specifics of a CRM CoE, it's essential to understand the terminology used within the CRM field, particularly in Salesforce contexts. A "CRM instance" refers to a singular, subscription-based unit that manages customer relationships, consisting of a primary production instance and related sandbox environments. In larger enterprises, we tailor multiple production instances to different business segments, integrating each with various platforms to support its business line. Sandboxes are environments for developing, testing, and deploying solutions to the production instances.

Understanding the management complexities of multiple production instances is crucial to determining the optimal structure for a CoE. The CoE harmonizes all CRM instances within a company, fostering enhanced business value through a unified approach.

Introducing the framework

This framework, rooted in industry best practices, offers a strategic road map to address current CoE challenges and foster continuous improvement of the CoE and its ability to derive business outcomes from its CRMs. Toward that end, the book guides you through a structured process designed to mature the CRM Center of Excellence.

We start by introducing the Five-Step Process for CRM CoE refinement. It provides a straightforward, systematic approach to identify, refine, and enhance critical functions within an organization. Selecting specific functions for refinement, drilling down to their current performance, defining ideal standards, executing targeted improvements, and validating the outcomes ensures that the CoE evolves in alignment with strategic business goals. This structured progression addresses immediate operational needs and lays the foundation for long-term success and adaptability in a dynamic business environment.

The following chapters will delve deeper into the framework's structure. This chapter presents an overview of the CRM CoE's realm of interest, domains, capabilities, and functions. Each component is crucial in driving excellence and achieving business outcomes, such as growing revenue, increasing customer satisfaction, reducing operational expenses, improving user adoption, and lowering risk. By understanding and implementing this comprehensive framework, organizations will be well-equipped to navigate the complexities of CRM management and inspired and motivated to unlock its full potential, leading to significant positive changes in customer-centric business.

The CRM CoE realm of interest

The CRM CoE realm of interest is organized from general, overarching concepts to progressively more specific subsets. We call these, from most general to most specific, domains, capabilities, and functions.

The CRM CoE realm of interest comprises three domains: customer centricity, operations, and foundations. Each domain is further segmented into three capabilities, each broken down into three functions. This hierarchical structure facilitates targeted analysis and strategic alignment throughout the CoE. Throughout the book, we return over and over again, to explore the domains, capabilities, and functions. We will use this piece of the framework as a lens to understand the scope of our CRM CoE realm of interest and the work that needs to be done within it to achieve desired business outcomes.

The Five-Step Process for CRM CoE Refinement

Below are the five steps of the process for CRM CoE refinement. Later in the book, we will define each step further and explain how to use this process. At this point, we are merely introducing some of the terms and concepts that make up this CRM CoE Framework.

1. **Select functions for refinement.**

    Assess the CoE's twenty-seven functions, focusing on those experienced as weak by a cross-functional group of assessors from across the CoE realm of interest. Identify the three weakest functions that impact the lowest weak rung of the business outcome ladder. These three become your current priority for improvement.

2. **Drill down into the selected functions.**

    Conduct a comprehensive analysis of the current performance of the three prioritized functions. This analysis will help you pinpoint specific weaknesses of each function and inform the challenges and opportunities you identified for your targeted enhancement plans.

3. **Define ideals.**

    Establish a robust framework of principles, best practices, standards, and KPIs, drawing on resources like the tables in Chapter 2 for guidance. Customize to reflect business context, ensuring relevance and effectiveness.

4. **Execute improvement.**

    Based on the newly defined ideals, develop a detailed enhancement strategy for the chosen functions, firmly grounded in concrete business outcomes. The plan will involve changes to some combination of process, technology, and talent. The alignment of defined ideals, executed improvements, and desired business outcomes is vital to the success of the plan. Use Chapter 3 for guidance on aligning functions with strategic business outcomes and Chapter 4 as a reference for economic justifications. Secure necessary funding and people, assign clear accountability, and execute the plan.

5. **Validate and iterate.**

    Implement a cyclical process for reviewing strategy effectiveness and KPI relevance, adjusting based on tangible data and evolving goals. Engage with CoE stakeholders to align experiences with KPI trends and refine the improvement constructs accordingly. Reassess the targeted functions and determine whether to make another iteration of improvements to one or more of these prioritized functions, or to the prioritize another set of functions and to iterate through the Five-Step Process with that next set of functions.

By rigorously following this process, the CoE assures that it responds to immediate operational challenges and evolves, driving continuous progress and delivering value. We address the Five-Step Process again at the end of Chapter 3.

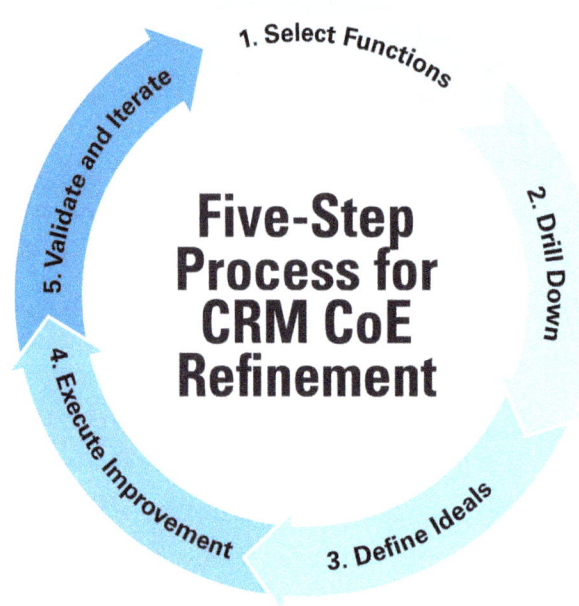

Figure 2: Five-Step Process for CRM CoE improvement and refinement.

CRM CoE Domains

Three domains comprise the CRM CoE realm of interest: customer centricity, operations, and foundation.

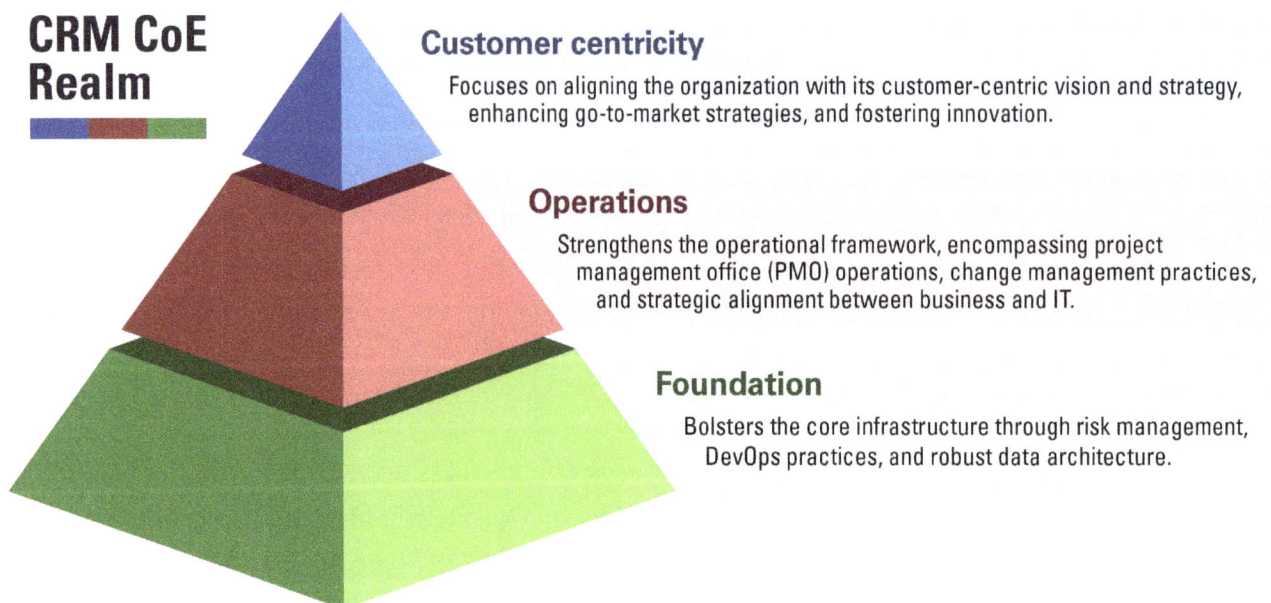

Figure 3: Domains in the CRM CoE realm of interest.

Capabilities

Within each domain, there are three specific capabilities. Each delivers organizational value through specialized expertise.

Customer-centric domain capabilities:

- **Vision and strategy** focus on developing clear strategic objectives for CRM initiatives that align with the broader organizational goals.

- **Go-to-market (GTM)** brings strategic plans to life in the marketplace by successfully implementing marketing, sales, and customer engagement strategies.

- **Innovation** drives the pursuit of new methods and products to keep the organization ahead of the curve.

Operations domain capabilities:

- **PMO** ensures efficient resource allocation and timely project delivery.

- **Change management** navigates organizational changes smoothly, providing effective transitions in processes, technology, and personnel.

- **Alignment** guarantees that business and IT units work cohesively towards shared goals.

Foundations domain capabilities:

- **DevOps** integrates software development with IT operations to enhance CRM system performance.

- **Risk management** identifies and mitigates risks to maintain operational integrity.

- **Data architecture** ensures strategic data management, design, and optimization.

CRM CoE Realm of Interest

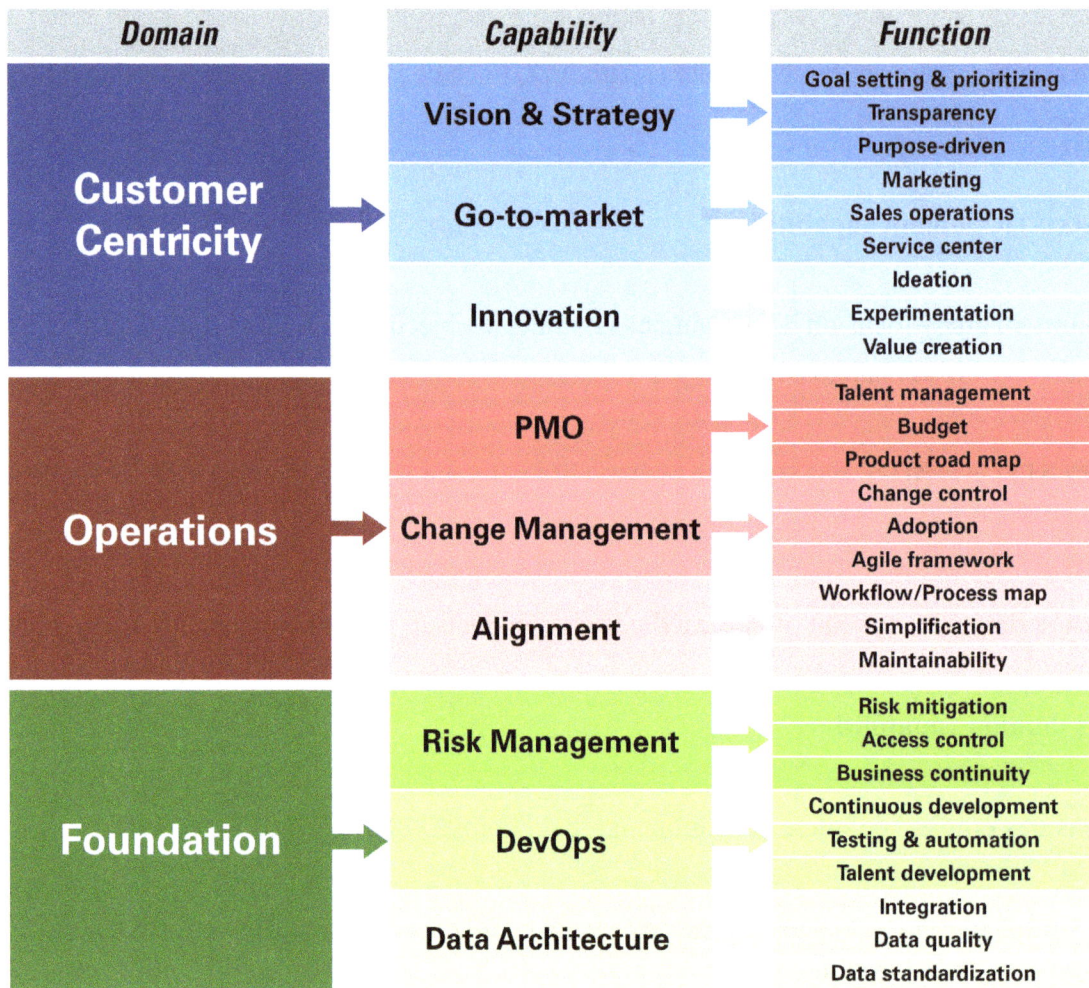

Domain	Capability	Function
Customer Centricity	**Vision & Strategy**	Goal setting & prioritizing
		Transparency
		Purpose-driven
	Go-to-market	Marketing
		Sales operations
		Service center
	Innovation	Ideation
		Experimentation
		Value creation
Operations	**PMO**	Talent management
		Budget
		Product road map
	Change Management	Change control
		Adoption
		Agile framework
	Alignment	Workflow/Process map
		Simplification
		Maintainability
Foundation	**Risk Management**	Risk mitigation
		Access control
		Business continuity
	DevOps	Continuous development
		Testing & automation
		Talent development
	Data Architecture	Integration
		Data quality
		Data standardization

Figure 4: The CRM CoE realm of interest is organized as domains, capabilities, and functions.

Functions

Each capability comprises three functions, the actionable tasks that drive the CoE towards achieving its strategic goals. These range from the goal-setting and prioritizing function, which defines and communicates the CoE's objectives, to the transparency function in CRM projects and more technical functions like continuous development and data integration. It is essentially at the function level that we target CRM CoE excellence and add tangible business value.

We transform functions into defined actionable workstreams that can be leveraged effectively, each with specific improvement criteria. Though teams or individuals are assigned accountability or responsibility for a function, there is often not a one-to-one relationship between a function and a team. Functions are often addressed across teams. The functions are synergistic, meaning improvement in one function can enhance others. The choice of functions to prioritize for immediate enhancement is strategic. Later, we'll discuss the rationale behind these choices and the methodology used to identify subsequent targets.

Cultural Aspects

Identifying where an organization can improve requires examining its vertical and horizontal dimensions. This framework's vertical dimension is structured as domains, capabilities, and functions, each building upon the lower layer(s). Conversely, CoE Cultural Aspects represent the horizontal dimension, encompassing critical subsets of functions that interact across different capabilities and domains. The CRM CoE Cultural Aspects of data, people, processes, technology, and agility are each composed of cross-domain and cross-capability sets of interconnected supporting functions. When these Cultural Aspect sets of functions are viewed and enhanced as a set, the organization's cohesion and operational sustainability improves.

Data
Ensures that all decision-making and strategic actions are backed by accurate, timely, and useful information.

People
Focuses on the development and engagement of personnel, fostering a culture that supports continuous improvement and innovation.

Process
Involves refining and optimizing workflows to increase efficiency and effectiveness.

Technology
Emphasizes adopting and integrating advanced technologies to enhance CRM capabilities.

Agility
Encourages flexibility and responsiveness to changes in the market or organizational environment, allowing the CoE to adapt quickly to new challenges.

Figure 5: CoE Cultural Aspects that affect sustainable refinement of the process.

Employing the CoE Cultural Aspects lens as a guiding framework enriches the maturity process of specific functions while uplifting the whole. This approach involves identifying the weakest cultural aspect and, when improving a function within that cultural aspect, determining how the other functions in that cultural aspect subset can also be touched at the same time. This method not only strengthens individual functions but also fosters a sense of unity and collaboration across seemingly disparate areas, leading to a more integrated and effective organizational matrix. Chapter 3 will delve deeper into the significance of CoE Cultural Aspects and how to use them to drive transformation across functions.

Business outcomes

The CRM CoE Framework adopts a results-oriented perspective, organizing functions into a matrix that supports the achievement of **five critical business outcomes**. Think of these as rungs on a ladder, starting with the most fundamental and progressing upward.

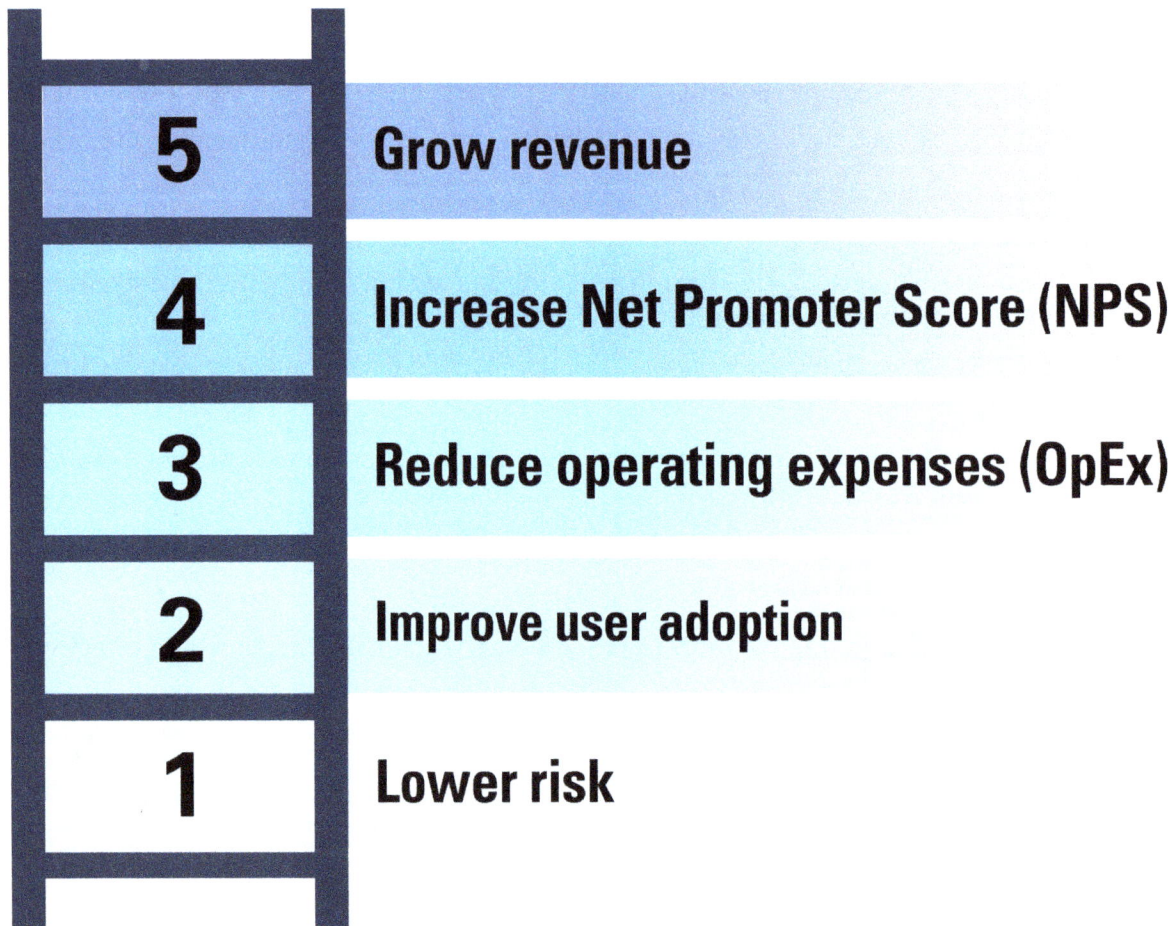

Figure 6: The business outcomes ladder.

Climbing the business outcomes ladder requires addressing the lower rungs first and then moving up the ladder.

1. **Risk mitigation.** Effectively managing and minimizing risks that endanger the CRM system and, in turn, the company is essential for risk reduction.

2. **Enhanced user adoption.** Increasing the utilization and involvement of CRM systems by users, including active participation from senior management.

3. **Optimized operational expenses.** Reducing and streamlining costs within the CRM ecosystem is crucial.

4. **Improved Net Promoter Score (NPS).** Enhancing customer satisfaction and loyalty metrics is crucial for boosting the NPS.

5. **Revenue growth.** The revenue growth outcome aligns CoE activities with growth strategies, measuring the impact on revenue through metrics, such as increasing sales, gaining market share, and optimizing pricing.

This sequential bottom-up approach ensures that enhancements in CoE performance begin with foundational outcomes before advancing to higher-level objectives. This prioritization strategy directs the focus toward functions crucial for achieving the baseline business outcomes, thereby ensuring a structured and sustainable improvement trajectory.

Conclusion

Chapter 1 laid the groundwork through terms and concepts for a comprehensive understanding and implementation of the CRM CoE Framework. This framework is designed to optimize customer relationship management initiatives and align them with business objectives, ultimately improving business outcomes and ensuring long-term success. The framework's ability to ensure consistency, facilitate benchmarking, and enhance organizational collaboration by providing a common language and standardized practices is critical to its success.

Acknowledging the variability in CoE operations, the chapter emphasizes the need for a balanced approach that leverages formal and informal influences. Understanding the complexities of managing multiple CRM instances is crucial to determining the optimal structure for a CoE. The Five-Step Process for CRM CoE refinement offers a practical road map for continuous improvement and strategic alignment.

As you delve deeper into the framework's structure in subsequent chapters, you will gain the tools and knowledge necessary to navigate the complexities of the journey toward CRM CoE maturity, which promises to unlock the full potential of CRM systems, leading to significant positive changes and sustained business success.

Insight Inquiry

How does the framework help manage multiple CRM environments?

As a practical tool, the framework plays a crucial role in harmonizing multiple CRM environments. By standardizing terminology, practices, and principles, it ensures consistency and efficiency, reduces complexity, and facilitates better governance. This structured approach is particularly beneficial for companies with multiple CRM production instances, as it aligns efforts and streamlines management.

How can CoEs ensure continuous improvement in a dynamic business environment?

CoEs ensure continuous improvement in a dynamic business environment by conducting regular assessments, maintaining feedback loops, and adapting to emerging challenges and opportunities. The Five-Step Process provides the loop that ensures you are identifying in each iteration the functions that emerge as those most in need of enhancement. The framework encourages ongoing learning and adaptation by integrating new insights and feedback into existing processes. This adaptive approach ensures the CoE remains relevant and effective in a rapidly changing business landscape, always aligned with the latest business strategies and technological advancements. By engaging the framework's lens of the Cultural Aspects—data accuracy, people development, process optimization, technology integration, and organizational agility—the CoE's continuous improvements become more sustainable.

How many functions can you target at one time?

Given the broad range of responsibilities within a CRM CoE, it's crucial to prioritize improvement efforts. Tackling no more than three functions simultaneously allows for manageable, meaningful enhancements. These functions are prioritized based on their alignment with core business outcomes. Engaging lower-rung business outcomes first stabilizes the fundamental aspects of the CRM ecosystem, supporting the success of more advanced business objectives.

CHAPTER 2
The Cookbook

Executive summary

Chapter 2, The Cookbook, establishes foundational elements for strategically enhancing a customer relationship management (CRM) Center of Excellence (CoE). It aligns with step three of the five-step CRM CoE refinement process, defining ideals. As such, it is a valuable tool for CoE leaders, business strategists, and organizational decision-makers. This chapter outlines principles, best practices, standards, and KPIs to systematically refine each function within the CRM CoE realm of interest, aiming to improve business outcomes and long-term success by aligning customer relationship management with the organization's core mission and goals.

Given the scope of the twenty-seven functions listed in Figure 4, the chapter advocates improving one to three functions simultaneously to ensure manageable and meaningful progress. This strategy makes it easier for CoE leaders to implement and track improvements. Organizations are encouraged to revisit this section as they transition focus from one target function to another, gathering necessary constructs for the following functions' enhancement iteration.

The chapter underscores the importance of starting with principles to provide benchmarks for current operations and identify deviations from company policies. It highlights best practices as critical for aligning the CoE with industry standards and guiding it toward optimal business outcomes. Selecting and adjusting KPIs ensures that organizations can track progress and success effectively, tailoring metrics to evolving needs. Integrating automation tools supports implementing principles, simplifies KPI measurement, and reinforces adherence to best practices and standards.

This chapter is a comprehensive resource, equipping CoE leaders to define their ideals for each function that they target for improvement. By systematically applying the outlined principles, best practices, standards, and KPIs, organizations can drive efficiency, collaboration, and effectiveness across production instances, functions, and teams, locking in more sustainable business outcomes.

The Cookbook

Why the cooking analogy? This concept stems from Charlie's experience as a Culinary Institute of America graduate, where he learned about "mise en place," a French term that means "put in place" or "gathering." In culinary environments, it refers to having ingredients, cookware, and utensils ready and in place to begin work. This organizational sensibility is woven into this book, particularly in calling this chapter "The Cookbook" and discussing the function ideals as recipes.

Few of us read cookbooks cover to cover or follow every recipe strictly. As novice cooks, we follow the recipes precisely, but experienced cooks and chefs use a cookbook as a flexible reference, with recipes as inspiration or good starting points. This book and this chapter, in particular, act similarly: They provide excellent starting points for developing each capability and function within the CRM CoE realm of interest. All are designed to be adaptable to your specific needs and circumstances.

To understand this, we will now dive into high-level sample ideals for each function. What follows is organized by domain, the three capabilities of each domain, and the functions related to each capability. Each component of the framework has a brief definition. Functions are detailed with three principles, a best practice for each principle, and three standards per best practice. We then offer examples of each function's key performance indicators (KPIs). The following are the base starting recipes for governing each of the twenty-seven functions.

---

1. DOMAIN: CUSTOMER CENTRICITY

This domain places the customer at the core of all CRM-related decisions and initiatives, aiming to enhance customer experiences and satisfaction with and through CRM solutions.

Capability: Vision and strategy

Capability focuses on developing and articulating clear goals and strategic plans for CRM initiatives, aligning them with the company's overall mission and purpose.

Function 1: Goal setting and prioritizing[3]

This function sets company goals, prioritizes objectives, and communicates relevance, emphasizing the importance of establishing clear, prioritized CRM goals that are aligned with the organization's business value and driven by executive sponsorship and demand planning processes

Principle	Best Practice	Standard
Everyone in our company's CRM ecosystem can articulate how a CRM's goals address the corporate mission or purpose.	As part of onboarding, all new CRM team members learn about how CRM goals address the corporate mission. These goals and their links to corporate mission are visible in big charts and CRM team information radiators. The goals are embedded in CRM reports and other communications.	Senior business stakeholders prioritize goals for each CRM according to their business value. CRM-related customer-centric, operational, and functional stakeholders collaborate on goals for each CRM. A mechanism exists to communicate the prioritized goals of each CRM to all staff working in that CRM's ecosystem.
The work of the CoE and CRM is driven by a demand planning process. The business approves the resulting plan.	Develop a CRM solution-planning process for demand, guided by the CoE, that begins with business needs, targets business objectives, and includes plans approved by the business.	The CoE participates in the budget process to support prioritized goals and demand planning. Demand planning considers required functionality, user support, and planned data growth. It is clear to each CRM staff person how they contribute to achieving the CRM's prioritized goals.
Executive sponsorship for the CoE and each CRM instance exist.	Identify and designate executive sponsorship for the CoE and for each CRM instance with defined support, guidance, and advocacy checkpoints.	The CoE executive sponsor and leadership team are involved in setting and fully embracing the CoE's prioritized goals. The CoE executive sponsor understands the prioritized goals of each CRM. The executive sponsor for each CRM is involved in setting and fully embracing the prioritized goals of the CRMs.

Figure 7: Example of the goal setting and prioritizing function.

KPI examples for the goal setting and prioritizing function:

- **Goal achievement.** The CoE tracks the percentage of goals achieved on time.
- **Employee alignment.** Surveys measure employee understanding and alignment with CRM goals.
- **Prioritized goals.** The CoE monitors the number of prioritized goals achieved per quarter or year.
- **Time to implementation.** The CoE measures the time taken from goal ideation to implementation.
- **Stakeholder satisfaction.** Surveys measure stakeholder satisfaction with the goal prioritization process.

---

3   All domain, capability, and function colors are keyed to the three overarching domain colors in Figure 4, the CRM CoE Realm of Interest chart.

Function 2: Transparency[4]

This function focuses on clear communication and openness across the CoE realm of interest.

Principle	Best Practice	Standard
Business metrics are made visible and known to all relevant stakeholders	Consistent messaging across formal and informal communication channels exposes all stakeholders to relevant business metrics, from internal end users to executive sponsors to CoE and CRM teams. We make such metrics known through broad distribution via staff meetings and electronic communications, internal newsletters, and team information radiators.	We have a scheduled cadence for outcome metrics for each CRM instance and the metrics being made visible and known. Discussions occur across and within CRM teams and stakeholders regarding outcome metrics, successes, and deficits. CRM outcome metrics influence priorities for future CRM work.
Communication is open among the value-stream partners.	Implement regular and periodic meetings on the state and enhancement of the CRM ecosystem, among and across all levels and verticals of value-stream partners.	Regular and periodic communication channels exist across organizations, the CoE, and the value-stream partners. Discussions occur within CoE teams and with stakeholders regarding CoE outcome metrics, successes, and deficits. CoE outcome metrics influence priorities for future CoE work.
All relevant stakeholders know the planned and delivered work.	The CoE conducts regular and periodic meetings and generates communications so that all stakeholders can understand the planned work, the delivered work, and any known positive or negative impact from that delivered work.	The CoE informs all team members of the descriptions of the planned functionality. The CoE informs all team members of work-in-progress demos. The CoE informs all team members of delivered, aborted, and deferred work.

Figure 8: Example of the transparency function.

KPI examples for the transparency function:

- **Frequency of transparent communication.** The CoE updates stakeholders and holds meetings regularly, with agreed-upon cadence.
- **Stakeholder engagement levels.** The CoE tracks stakeholder engagement in CRM projects.
- **Access to CRM performance metrics.** The CoE ensures that CRM employees have access to CRM performance metrics.
- **Lessons-learned sessions.** The CoE conducts lessons-learned sessions at a specified frequency to review and improve processes.
- **CRM project delivery success rate.** The CoE monitors and reports on the success rate of CRM project deliveries.

---

4   Kim Scott, *Radical Candor: Be a Kick-Ass Boss Without Losing Your Humanity* (New York: St. Martin's Press, 2017). Darrell K. Rigby, Sarah Elk, and Steve Berez, *Doing Agile Right: Transformation Without Chaos* (Boston: Harvard Business Review Press, 2020).

Function 3: Purpose-driven[5]

This function emphasizes the alignment of CRM strategies with creating tangible business value.

Principle	Best Practice	Standard
Team members work on measurable deliverables aligned with business goals.	Assign deliverables associated with desired measurable outcomes aligned with business goals to team members' work queues. Inform the assigned team members of the deliverables, quantifiable outcomes, and related business goals. Upon delivery, make the team members aware of the outcomes.	All team members can name a current priority for generating corporate economic driver improvement from their related CRM. All team members can name a future priority for generating economic driver improvement from CRM. All team members can name a specific past corporate economic driver improvement generated from CRM.
Every CRM ecosystem staff person can name the core customer served by their CRM instance(s).	The CoE continuously articulates written descriptions of the CRM's core customer(s) within CRM ecosystem staff communications.	I can describe CRM's role in supporting corporate economic drivers. All my team members can describe CRM's role in supporting corporate economic drivers. All team members know where to find the metrics showing the impact of CRM's role in supporting corporate economic drivers.
Every CoE ecosystem staff member can name an idea for improving the CRM's impact on our business outcomes.	Solicit periodic ideas from throughout the CoE ecosystem for improving the CRM's impact on the business's economic drivers. Create a means to say thank you for all input and to acknowledge input that becomes the impetus for enhancements.	I know how to improve the impact on company economic drivers through CRMs. Others I work with would have ideas about how to improve the impact on the CRM's economic drivers. The CoE road map aligns with the company's economic drivers.

Figure 9: Example of the purpose-driven function.

KPI examples for the purpose-driven function:

- **Number of initiatives directly contributing to corporate economic drivers.** The CoE tracks the number of initiatives that contribute to the company's economic drivers.
- **Employee engagement in value-driven activities.** The CoE measures employee engagement through surveys.
- **Percentage of CRM ecosystem staff able to name and describe the CRM's core customer.** The CoE surveys the CRM ecosystem staff to determine the percentage who can articulate the core customer.
- **Number of ideas generated to improve economic drivers.** The CoE monitors the number of ideas generated to improve economic drivers.
- **Alignment of CoE road map with company economic drivers.** The CoE evaluates the alignment of the CoE road map with the company's economic drivers.

---

5   Eric Ries, *The Lean Startup: How Today's Entrepreneurs Use Continuous Innovation to Create Radically Successful Businesses* (New York: Crown Business, 2011).

Capability: Go-to-market

This capability within the customer centricity domain focuses on marketing, sales, and customer engagement strategies using the CRM.

Function 1: Marketing[6]

This business operations function focuses on outreach as supported by the CRM.

Principle	Best Practice	Standard
The company's market channels are clearly defined.	Work with relevant stakeholders to define marketing channels, their priorities, purposes, methods, and people responsible for them, and how each is related to a CRM instance.	I can name at least one of our market and customer channels served by the CRM system. I can describe our engagement strategy for that market and customer channel and how that strategy uses the CRM system. Others I work with can also identify market and customer channels served by the CRM and the related engagement strategies.
The relevant stakeholders understand the company's marketing objectives well.	Work with relevant stakeholders to define marketing objectives, their priorities, purposes, methods, and people responsible for them, and how each is related to a CRM instance.	The ways in which our CRM system serves our market and customer channels is defined. I know the priority of each of the channels served by the CRM system. Others I work with know the priorities of the market and customer channels served by the CRM system.
The company's marketing and sales teams are aligned.	Foster collaboration for marketing and sales teams to align marketing channels and objectives with sales strategies, operations, practices, CRM capabilities, and road map.	Each of my teammates can describe a channel's engagement strategy, how the CRM supports that strategy, and its expected outcomes. Each of my teammates can describe a channel's engagement strategy, how the CRM supports it, and its expected outcomes. My teammates can name a metric used to measure a channel engagement strategy supported by the CRM.

Figure 10: Example of the marketing function.

KPI examples for the marketing function:

- **Number of defined marketing channels effectively using CRM.** The CoE tracks the number of marketing channels that utilize the CRM system.
- **Rate of achievement of CRM-dependent marketing objectives.** The CoE measures the rate at which marketing objectives dependent on the CRM are achieved.
- **Alignment score between marketing and sales strategies.** The CoE assesses the alignment score between marketing and sales strategies.

---

6   W. Chan Kim and Renée Mauborgne, *Blue Ocean Strategy: How to Create Uncontested Market Space and Make the Competition Irrelevant*, expanded ed. (Boston: Harvard Business Review Press, 2015).
Philip Kotler and Kevin Lane Keller, *Marketing Management*, 15th ed. (Boston: Pearson, 2016).

- **Revenue growth attributed to CRM-driven marketing strategies.** The CoE monitors the revenue growth attributed to CRM-driven marketing strategies.
- **Customer engagement and conversion rates from CRM-led campaigns.** The CoE evaluates the customer engagement and conversion rates resulting from CRM-led marketing campaigns.

Function 2: Sales operations[7]

This business operations function is related to sales processes supported by the CRM.

Principle	Best Practice	Standard
The company's sales strategy is clearly defined.	Work with relevant stakeholders to define sales strategies, their priorities, purposes, methods, and people responsible for them, and how each is related to a CRM instance.	CRM sales data per channel engagement strategy informs sales-enabling decisions. The CRM ecosystem data is beneficial to understanding the effectiveness of sales channel strategies. Optimizing CRM to better support the sales operations and enabling priorities is part of the CRM road map discussions.
The company's sales operations are more effective when leveraging the CRM platform.	Identify the gaps in the sales operation's ability to leverage the CRM platform and create a road map for implementing enhancements.	The CRM processes enable our sales methodologies. The CRM reports show progress toward meeting sales goals. We build CRM solutions to help meet sales goals.
We reflect the company's sales best practices in the CRM road map.	Review the company's sales best practices, determine gaps in alignment with the CRM road map, and create a plan for remediating the CRM road map.	Sales operations and CRM team maintain planning communications to improve CRM's support of sales operations. Sales operations maintain a list of desired improvements to the CRM. CRM team builds and tests solutions with active input from sales operations. Sales operations feedback on CRM performance and enhancement needs is systematically gathered and acted upon.

Figure 11: Example of the sales operations function.

KPI examples for the sales operations function:

- **Sales strategy implementation success rate.** The CoE measures how effectively the sales strategy is implemented in the CRM.
- **CRM's impact on sales operations efficiency.** The CoE tracks the efficiency improvements in sales operations resulting from CRM utilization.
- **Number of sales goals met or exceeded through CRM utilization.** The CoE monitors the number of sales goals that are met or exceeded due to CRM use.
- **Rate of improvement in the sales process due to CRM adoption.** The CoE measures the rate of improvement in the sales process due to adopting the CRM.

---

7   Neil Rackham, *SPIN Selling* (New York: McGraw-Hill, 1988).

- **Alignment of sales best practices with CRM road map.** The CoE evaluates how well the sales best practices align with the CRM road map.

Function 3: Service center[8]

This business operations function is related to service processes supported by CRM.

Principle	Best Practice	Standard
The company's service center has a 360-degree view of the customer.	Design the architecture of the service center's CRM instance to provide a 360-degree view of each customer being served.	Each person who works with a CRM instance in the service center can name several ways in which the CRM system serves the corporate mission. Each person who works with the CRM in the service center can name the service center metrics that the CRM is measuring.
There are clearly defined measurable outcomes for the CRM's role in serving the service center.	The service center aims to understand desired measurable outcomes, clarify and identify definitions and appropriate sources of information, and create means within the CRM to measure those metrics.	The service center utilizes the CRM's service metrics to evaluate its success in meeting corporate mission goals. The customer's perspective on the service center's performance is reflected in CRM-reported metrics. Service center KPI metrics generated by the CRM are visible for all working in and supporting the service center.
The service center has a list of desired improvements to the CRM's ability to address the service center's goals.	Conduct periodic workshops to create a backlog of improvements with priority rankings for the service center's business needs.	Service center and CRM teams maintain planning communications to improve CRM's support of the service center. The service center team maintains a list of desired improvements to the CRM. The CRM team builds and tests solutions with active input from the service center.

Figure 12: Example of the service center function.

KPI examples for the service center function:

- **Customer satisfaction score for CRM-driven services.** The CoE tracks customer satisfaction scores for the services it facilitates.
- **Number of service center goals achieved through CRM.** The CoE monitors the number of service center goals achieved through the CRM system.
- **Service center response and resolution time.** The CoE measures the service center's response and resolution time using CRM data.
- **Success rate of 360-degree customer-view implementation.** The CoE evaluates the success of implementing a 360-degree view of customers within the service center.
- **Rate of service center improvement in backlog clearance.** The CoE tracks the rate at which the service center clears its improvement backlog using CRM solutions.

---

8   John A. Goodman, *Customer Experience 3.0: High-Profit Strategies in the Age of Techno Service* (New York: AMACOM, 2014).

Capability: Innovation

This capability evaluates and implements new methods or products.

Function 1: Ideation[9]

This function engages employees in developing new ideas for products and services.

Principle	Best Practice	Standard
The company has a formal process to capture ideas from its relevant stakeholders, including external sources.	Foster collaboration to generate ideas from relevant stakeholders, including external sources, and establish processes for a more effective continuous collection of ideas to improve the CRM ecosystem.	When brainstorming for and prototyping improvements to business, CRM team members are included from across the supporting ecosystem. When designing CRM solutions for sticky problems, we consult our CRM vendor to see how they have solved similar issues elsewhere. When CRM teams determine the problem solutions, business representatives are consulted for validation.
The resulting solution from innovative ideas syncs with the company's larger innovation ecosystem.	Build an ideation process that funnels CRM solutions into an innovation ecosystem, then aligns and prioritizes them against the larger corporate goals.	When designing CRM solutions for sticky problems, we consult our CRM vendor to see how they have solved similar issues elsewhere. When the CRM's limitations prohibit a practical, efficient, or economical solution, we talk to our CRM vendor requesting consideration in their future road map. We periodically meet with our CRM vendor to inform them of our road map and needs.
The organization has a formal process to implement approved solutions from the innovation ecosystem.	Define and implement a formal ideation process for approving, funding, and ushering approved solutions to completion.	Our CRM team stays abreast of our CRM's coming functionality releases. Our CRM team tests coming CRM functionality in our sandboxes to anticipate issues and take advantage of new solutions. Our CRM team actively aligns our road map with our vendor and works to have our vendors' releases enhance our own.

Figure 13: Example of the ideation function.

KPI examples for the ideation function:

- **Number of ideas generated through the CRM ecosystem**. The CoE tracks the number of new ideas generated within the CRM ecosystem.
- **The conversion rate of ideas into actionable CRM improvements**. The CoE measures the rate at which generated ideas are converted into actionable CRM improvements.
- **Alignment of innovative ideas with larger organizational goals**. The CoE evaluates how well innovative ideas align with larger organizational goals.
- **Stakeholder participation in the ideation process**. The CoE monitors stakeholder participation in the ideation process to ensure broad engagement.
- **Impact of implemented ideas on NPS or customer satisfaction**. The CoE assesses the effects of implemented ideas on the Net Promoter Score (NPS) or overall customer satisfaction

---

9   Clayton M. Christensen, *The Innovator's Dilemma: When New Technologies Cause Great Firms to Fail* (Boston: Harvard Business Review Press, 1997).

Function 2: Experimentation[10]

This function develops a culture of testing new ideas through experimentation.

Principle	Best Practice	Standard
Fail-fast experiments[11] and hypothesis-based testing of new technologies, processes, and tools are encouraged.	Build processes and approvals for quick response to proposed fail-fast experiments and hypothesis-based testing of new technologies, processes, and tools, expediting approval at the lowest level possible based on risk, time, and cost.	Our budget process and staffing pattern allow time for CRM-related experiments. A process exists to generate recommendations for experiments seeking solutions to CRM issues. CRM-related experiments have been initiated this last year.
Organization involves multiple teams to collaboratively experiment continually on new approaches.	We endorse managers quickly composing a cross-team, ad hoc group to experiment on a new approach to determine its viability collaboratively.	CRM experiments are designed to learn something useful and iterate upon that learning. CRM experiments are limited in duration. CRM experiments are run with three possible outcomes: 1) fail, learn, abort, 2) fail, learn, pivot, 3) succeed and learn.
The organization has a formal documentation process to capture experiment outcomes.	Create and disseminate an organizational repository capturing the scope and results of experimented approaches and outcomes.	Someone owns each experiment and the decisions about when to abort, pivot, or implement. Records of experiments are kept so that we can reuse pieces of them if they later prove useful for another purpose or iteration. Whoever conducted the experiment performs a retrospective after completing the experiment to collect insights into what worked and what did not.

Figure 14: Example of the experimentation function.

KPI examples for the experimentation function:

- **Number of fail-fast experiments conducted.** The CoE tracks the number of fail-fast experiments conducted.
- **The success rate of collaborative CRM experiments.** The CoE measures the success rate of collaborative CRM experiments.
- **Documentation and utilization rate of experiment outcomes.** The CoE evaluates how well experiment outcomes are documented and utilized.
- **Percentage of experiments leading to tangible CRM improvements.** The CoE monitors the rate of experiments that result in tangible CRM improvements.
- **Return on investment for CRM-related experiments.** The CoE assesses the return on investment for CRM-related experiments.

---

10   Jeff Gothelf and Josh Seiden, *Lean UX: Applying Lean Principles to Improve User Experience* (Sebastopol, CA: O'Reilly Media, 2013).
Safi Bahcall, *Loonshots: How to Nurture the Crazy Ideas That Win Wars, Cure Diseases, and Transform Industries*, (New York: St. Martin's Press, 2019).
11   Ryan Babineaux and John Krumboltz, *Fail Fast, Fail Often: How Losing Can Help You Win* (New York: TarcherPerigee, 2013).

Function 3: Value Creation[12]

This function develops new products, services, or approaches for internal or external business value.

Principle	Best Practice	Standard
The organization has a team that promotes value creation.	Assign specific people and teams the accountability to promote creating value through CRM solutions.	Sandbox environments are set aside for prototyping outside the release development cycle. Teams or individuals can request time to prototype an idea. Prototype efforts are limited in duration.
The organization has a formal funding process for value-creation efforts.	Define a formal process for obtaining funding to implement a value-creation effort.	Prototypes are solutions built by a cross-functional team. Prototyping is expected to consider AppExchange and declarative solutions as part of the solution mix. The prototyping team is expected to answer whether this solution will reduce system friction or add value to customers.
The organization develops products or services from value creation that can be sold to customers.	Require each value-creation effort to identify and report whether it generates new products or services that can be sold to customers. For those representing such an opportunity, define a team and process to evaluate whether it is worthwhile.	Prototypes are documented so that pieces that may later prove helpful can easily be retrieved. The prototype is designed to get stakeholder feedback, but someone owns the prototype go or no-go decisions. A lesson-learned session follows creation of the prototype to glean and document what was learned.

Figure 15: Example of the value creation function.

KPI examples for the value creation function:

- **Number of new products or services developed from CRM initiatives.** The CoE tracks the number of new products or services developed from CRM initiatives.
- **Revenue generated from CRM-driven value-creation efforts.** The CoE measures revenue generated from CRM-driven value-creation efforts.
- **Rate of successful prototype development and implementation.** The CoE monitors the rate of successful prototype development and implementation.
- **Stakeholder feedback on CRM-driven value creation.** The CoE collects and evaluates stakeholder feedback on CRM-driven value creation.
- **Funding efficiency for value-creation projects.** The CoE assesses the efficiency of funding for value-creation projects.

---

12   Clayton M. Christensen and Michael E. Raynor, *The Innovator's Solution: Creating and Sustaining Successful Growth* (Boston: Harvard Business Review Press, 2003).
Alexander Osterwalder and Yves Pigneur, *Value Proposition Design: How to Create Products and Services Customers Want* (Hoboken, NJ: Wiley, 2014).
Nir Eyal, *Hooked: How to Build Habit-Forming Products* (New York: Portfolio, 2014).

2. DOMAIN: OPERATIONS

This domain covers strategies and practices for managing operational aspects of CRM implementations.

Capability: Project management office (PMO)

This capability focuses on talent allocation, budget, product delivery, and managing CRM projects.

Function 1: Talent management[13]

Our company's inside and outside talent allocations are aligned to tasks.

Principle	Best Practice	Standard
The organization has the right talent to deliver the solution.	Define 1) whether the organization lacks the right talent to deliver solutions in light of external or internal circumstances, 2) whether those talents are best kept inside the organization or hired in from contractors, and 3) whether that talent can be grown internally. Create a correction plan based on the conclusions.	When outside contractors build a CRM project, we build in-house skills to maintain and develop the completed solution. When outside contractors build a CRM solution, we include robust documentation of the solution in the contract. When our employees build a CRM solution, we augment their skills with outside contractors.
The organization can find and retain relevant talent.	Determine the internal and external circumstances that make finding and retaining relevant talent difficult, determine whether an opportunity exists to grow and promote internally, define parameters for when to contract for needed talent and when not, and create a talent remediation plan.	Our CRM employees are encouraged to develop their skill set, and money is available for them to do so. Our CRM employees have opportunities to work cross-functionally, learn new roles, and move up within the company. Our CRM employees have opportunities to switch between professional or managerial roles within their careers.
The teams collaboratively work to build timely solutions.	Determine the processes, structures, and disincentives that inhibit cross-team collaborative work and completing a solution on time. Create and implement a plan to remove blockages and promote cross-team collaboration.	CRM employees are eager to stay with us and build their careers in this company. New CRM talent is eager to work here. Our CRM employees actively refer others to work here.

Figure 16: Example of the talent management function.

KPI examples for the talent management function:

- **Employee skill enhancement rate post-CRM training.** The CoE tracks how employee skills improve after CRM training.
- **Retention rate of CRM-trained talent.** The CoE measures the retention rate of employees who have received CRM training.

---

13   Marcus Buckingham and Curt Coffman, *First, Break All the Rules: What the World's Greatest Managers Do Differently* (New York: Simon & Schuster, 1999).
Antonio Nieto-Rodriguez, *The Project Revolution: How to Succeed in a Project-Driven World* (London: LID Publishing, 2019).

- **Time to fill CRM-related positions.** The CoE monitors the time taken to fill CRM-related positions.
- **Employee satisfaction with CRM career development opportunities**. Through surveys, the CoE evaluates employee satisfaction with career development opportunities in CRM.
- **Effectiveness of cross-functional CRM collaborations.** The CoE assesses the effectiveness of cross-functional CRM collaborations.

Function 2: Budget[14]

This function manages the budget for CoE deliverables.

Principle	Best Practice	Standard
Our budget for CRM development is set annually.	Move to a rolling quarterly budgeting process within an annual intended spend, allowing changing business needs and project success. This is intended to influence allocated effort and budget in future quarters.	Our CRM is more than simply a cost center; it is also seen as a revenue generator. We use our CRM to segment customer interactions and contextualize and personalize customer messages. Our CRM is focused on driving higher sales conversion rates.
The organization periodically measures its CRM-derived ROI.	Conduct a workshop and design process for determining ROI on CRM solutions before building a solution and, where possible, incorporating the ROI metrics within the solution implementation.	Our annual budget for CRM development is based on estimated return on investment. Our annual budget for CRM maintenance is based on estimated return on investment. Our CRM development budget is based on the cost of the new functionality we wish to add each budget cycle.
Our organization prioritizes and balances funding among tech debt, maintenance, and new capabilities.	To utilize or accommodate the platform's new functionality, build into CRM product management team budgets an allocation of 20%–30% for updating code to the most recent CRM vendor code versions and security guidelines, redesigning the architecture, refactoring, and rewriting code. To incorporate very large enhancements from the CRM vendor's new releases, a separate non-project-based allocation may also be necessary.	Our budget for CRM maintenance assumes periodic changes in the platform due to CRM platform architecture and functionality changes. Our budget for CRM maintenance is tied to the number of people licensed and supported on the CRM platform. Changes in the number of licensed users, data storage growth, and the impact on backup costs are part of our budget management planning.

Figure 17: Example of the budget function.

KPI examples for the budget function:
- **ROI from CRM development projects.** The CoE measures the return on investment from CRM development projects.
- **Percentage adherence to CRM development budget.** The CoE tracks how closely the actual spending aligns with the budgeted amount for CRM development.

---

14   Glen S. Gooding, *The IT Financial Management Lifecycle: Budgeting, Costing, Chargeback, and Benchmarking* (London: Kogan Page, 2010).
Greg Horine, *Project Management Absolute Beginner's Guide*, 4th ed. (Indianapolis: Que Publishing, 2017).

- **Efficiency of the rolling quarterly budgeting process.** The CoE evaluates the efficiency of the rolling quarterly budgeting process in accommodating changing business needs.
- **Cost savings from optimized CRM maintenance.** The CoE measures the cost savings achieved through optimized CRM maintenance practices.
- **Alignment of budget with CRM platform changes.** The CoE ensures the budget aligns with changes and updates in the CRM platform.

Function 3: Product road map[15]

This function creates and manages a timeline for delivering business functionalities and platform capabilities.

Principle	Best Practice	Standard
The organization follows a road map of the planning and delivery process.	Implement an annual road map of the planning and delivery process with at least quarterly checkpoints and adjustments. Develop a team responsible for planning, checkpoints, and approving modifications and their relationship to the implementation team(s).	We develop a CRM release map and schedule, and we can modify it in response to changing business priorities. Our CRM release plan and long-term road map are derived from business and information technology collaboration. They involve participation from the three CoE domains: customer centricity, operations, and functions. CRM release maps and schedules are updated regularly.
The organization has a process for collecting and incorporating product release feedback into the future planning cycle.	Create a solution feedback process to incorporate needed changes into the future planning cycle.	CRM works with business to identify all features or services that require training and to plan for them. Release features that require user training can be deployed to production, but enabling can be delayed and switched on after users are trained to use the new features. When new features or services are utilized, we track their usage, impact, and end-user experience.
Backlogged projects, enhancements, and bug fixes are prioritized based on customer value, speed, and adaptability.	Determine and implement a prioritization process for backlogged projects, enhancements, and bug fixes based on three criteria: customer value, speed or ease of implementation, and business buy-in.	Each of our prioritized backlogged projects has at least a cost estimate and an estimated ROI assigned. Our processes and tools allow an emergency bug fix deployment while maintaining testing and traceability. Enhancements are prioritized in collaboration with the business owner.

Figure 18: Example of the product road map function.

KPI examples for the product road map function:

- **Alignment of CRM road map with business priorities.** The CoE assesses the road map alignment with the company's business priorities.
- **The success rate of product releases against the road map.** The CoE tracks the success rate of product releases against those scheduled on the road map.

---

15   C. Todd Lombardo, Bruce McCarthy, Evan Ryan, and Michael Connors, *Product Road Maps Relaunched: How to Set Direction While Embracing Uncertainty* (Sebastopol, CA: O'Reilly Media, 2017).

- **Stakeholder satisfaction with CRM product development.** The CoE measures stakeholder satisfaction with CRM product development through surveys and feedback.
- **Time to market for new CRM features.** The CoE monitors the time it takes to bring new CRM features to market.
- **The backlog clearance rate for CRM enhancements and fixes.** The CoE evaluates the rate at which backlogged CRM enhancements and bug fixes are addressed and resolved.

Capability: Change management

This is the capability for managing the adoption of change in processes, technology, and people.

Function 1: Change control[16]

This function introduces CRM changes systematically and safely.

Principle	Best Practice	Standard
The organization has defined change control standards, procedures, and processes.	Define change-control standards, procedures, and processes. Articulate rationales for each that reduce risk and allow solutions to move to production quickly. Define emergencies, how they are declared, what steps are taken, and who approves such exceptions. Create a plan for automating processes.	Roles assigned to ensure change-control processes are followed. No change occurs to production without approval by change control person/s or automated processes. We have defined change-control processes for handling an emergency change to production.
The organization has continuous education dedicated to change control processes.	Develop ongoing change-control training for all development roles.	All CRM solutions begin with a user story in a lower environment and are promoted to production following testing. Before a solution goes live in the CRM, the final tester represents business. Business approves the story to be built.
The organization keeps its change control process reviewed and updated.	Establish a periodic and regular review of change-control standards, procedures, and processes with steps for requesting emergency consideration of exceptions to standards, procedures, and processes. When such exceptional situations occur, include follow-up updates to the standards, procedures, and processes.	System integration partners may contract to work to more stringent standards but are never allowed lower standards than ours. System integration partners must maintain traceability and work within our DevOps, testing, and quality-control tool sets. Our change-control checks and balances apply to the system-integration partners who build our CRM solutions.

Figure 19: Example of the change control function.

KPI examples for the change control function:

- **The compliance rate with change-control standards.** To ensure adherence and consistency, the CoE monitors the compliance rate with change-control standards.
- **Time to successfully implement CRM changes.** The CoE measures the time from idea to successful CRM implementation.
- **Number of emergency changes and their impact.** The CoE tracks the number of emergency changes and assesses their impact on the CRM system and business operations.
- **Effectiveness of change-control training initiatives.** The CoE evaluates the efficacy of change-control training initiatives by measuring process adherence and implementation improvements.

---

16   John P. Kotter, *Leading Change* (Boston: Harvard Business Review Press, 1996).

- **ROI from CRM changes implemented.** To determine their value and contribution to business goals, the CoE assesses the return on investment from CRM changes implemented.

Function 2: Adoption[17]

This function enables users to adopt CRM systems and changes.

Principle	Best Practice	Standard
Defined strategies are implemented for onboarding, training, and continued training of new CRM system end users.	Define and implement strategies for onboarding, training, and continued training of new CRM system end users.	Training and end-user documentation are considered for each built solution. Just-in-time training is available to new users and for new features. CRM systems are built or integrated with other systems to automatically onboard users to CRM or vice versa.
End users are continually engaged to understand what will make the CRM system more efficient and faster for them to use.	Create regular and periodic processes for soliciting input from end users about what will make the CRM system more accessible and faster for them to use. Consider identifying "super users" who can help train newer users, assist in initial troubleshooting for end users, and act as a conduit of ideas aimed at enhancing the system for end users.	Adoption campaigns are in place to drive knowledge of the system and prompt users to engage the CRM. End user representatives define and test the solutions they will be trained to use. We can use a training sandbox for complex training or onboarding scenarios.
User adoption metrics are regularly reported across the ecosystem and evaluated for end-user experience improvements.	Conduct a workshop to identify the critical end-user metrics and report them across the ecosystem. Regularly evaluate for updates and use them in the enhancement feedback loop.	CRM solutions are built to enable users to do their jobs better and more efficiently. CoE and business work together to increase the user adoption of the CRM. A user license management strategy increases user adoption.

Figure 20: Example of adoption function.

KPI examples for the adoption function:
- **User adoption rate after CRM system changes.** To ensure effective implementation and utilization, the CoE measures user adoption following changes to the CRM system.
- **End user's satisfaction with CRM system usability.** To identify areas for improvement and enhance user experience, the CoE assesses end-user satisfaction with the system's usability.
- **Effectiveness of CRM onboarding and training programs.** The CoE evaluates the effectiveness of CRM onboarding and training programs by measuring user competency and engagement improvements.
- **Rate of successful CRM feature adoption.** To ensure users take advantage of system enhancements, the CoE tracks the rate at which new CRM features are adopted.
- **CRM license utilization and management effectiveness.** To optimize license allocation and usage and to ensure cost-effectiveness, the CoE monitors CRM license utilization and management.

---

17 Michael E. D. Koenig and David R. Koehler, *The Art of Training Delivery: How to Train Employees Effectively* (New York: Wiley, 2002).

Function 3: Agile framework[18]

This function is concerned with developing the Agile framework and its adoption by the organization.

Principle	Best Practice	Standard
The organization has a mature software delivery framework for its Agile development lifecycle.	Evaluate the Agile delivery framework for its ability to reduce cycle time, increase deployment frequency, increase throughput, and fix escaped defects. Identify the roadblocks to consistent delivery that positively impact those metrics and remediate the framework gaps.	Our Agile framework promotes, as stated in the Agile Manifesto, "satisfying the customer through early and continuous delivery of valuable software." All employees and contractors who develop and maintain the CRM systems receive training in our Agile software development lifecycle delivery framework and methodology. Teams are facilitated by people who have had scrum master training and support.
We have identified metrics to measure progress on our Agile journey.	Review the organization's ability to report metrics per cycle on the five DORA metrics (cycle time, deployment frequency, failures in production, time to restore service, and reliability). Determine whether other metrics are critical and include them.	Our processes encourage cross-functional work within the team and collaboration between businesspeople and developers. Our processes harness change for the customer's competitive advantage. Our processes encourage conveying information to and within a development team through conversation.
The organization can deliver incremental customer value frequently.	Determine the next steps in automation, processes, or an Agile framework to increase the delivery frequency of incremental customer value.	Our CoE teams track the number of releases, sprints, points, and bugs delivered for internal process review. CRM teams use delivery performance metrics like 1) deployment frequency, 2) cycle time, 3) time to recover, 4) throughput, and 5) escaped defects. CRM teams gauge value delivered through solution goal metrics defined by the business during work intake.

Figure 21: Example of the Agile framework function.

KPI examples for the Agile framework function:

- **Cycle time reduction due to Agile practices.** To ensure faster value delivery, the CoE measures the reduction in cycle time attributed to adopting and refining Agile practices.
- **Deployment frequency and success rate.** To assess the effectiveness of the Agile framework in facilitating continuous delivery, the CoE tracks deployment frequency and success rate.
- **Agile maturity-level assessment score.** To identify areas for improvement and ensure adherence to Agile principles, the CoE evaluates the maturity level of the Agile framework through periodic assessments.
- **Throughput and escaped defects metrics.** To gauge the efficiency and quality of the Agile development process, the CoE monitors throughput and escaped defects metrics.
- **Stakeholder satisfaction with Agile CRM delivery.** To ensure the framework meets business needs and expectations, the CoE measures stakeholder satisfaction with the delivery process.

---

18   Mike Cohn, *Agile Estimating and Planning* (Upper Saddle River, NJ: Prentice Hall, 2006).

Capability: Alignment

This capability aligns business and IT units on the value stream.

Function 1: Workflow/process map[19]

This function involves articulating, analyzing, documenting, and improving business processes and IT workflows.

Principle	Best Practice	Standard
The organization has defined a customer journey.	Conduct a cross-team workshop to define the customer journey, including the people, processes, CRM, and other technology engaged with customers at each point of their interaction with the organization. Determine gaps in that journey and how they might be remediated.	Our mapped process knowledge includes all the people, expertise, and content needed to execute a business process. Our CRM-assisted business processes push information to the right people or systems according to their execution roles. Our CRM-assisted business processes are well integrated with other sub-processes and add value to the entire process.
The organization follows a consistent delivery process.	Analyze the delivery process portion of the customer journey to determine where consistency would improve the delivery process. Determine whether modifying the steps within the CRM would make delivery more reliable.	Our CRM-assisted processes are designed to limit or remove siloed divisions between our organization's groups. CRM solution design considers company goals and not just local ones. We avoid tunnel vision in process mapping by having all the impacted groups at the table.
The organization leverages out-of-the-box CRM functionalities.	Assign accountability and responsibility for advance review of the CRM vendor's periodic functionality updates for new out-of-the-box functionalities to ease the user experience, reduce code that needs to be maintained, and make the system more reliable. Build all new enhancements with an out-of-the-box first mindset.	We map our processes across the customer journey and evaluate CRM solutions for their impact on the customer experience. Our CRM-assisted processes encourage information transfer between the people working to meet customer needs. Our data hub is centered on our customer experience.

Figure 22: Example of the workflow/process map function.

KPI examples for the workflow/process map function:

- **Efficiency improvement in CRM-supported customer journey.** The CoE measures the improvement in efficiency within the customer journey facilitated by CRM-supported processes.

-

---

19   John Jeston and Johan Nelis, *Business Process Management: Practical Guidelines to Successful Implementations*, 3rd ed. (London: Routledge, 2014).
Tristan Boutros and Tim Purdie, *The Process Improvement Handbook: A Blueprint for Managing Change and Increasing Organizational Performance* (New York: McGraw-Hill, 2013).
Quentin Brook, Lean Six Sigma and Minitab: *The Complete Toolbox Guide for Business Improvement*, 5th ed. (Winchester, UK: OPEX Resources, 2014).

- **Consistency and reliability of CRM-assisted delivery processes.** The CoE tracks the consistency and reliability of delivery processes enhanced by CRM assistance.
- **User experience improvement due to out-of-the-box CRM functionalities.** The CoE evaluates the improvement in user experience resulting from implementing out-of-the-box CRM functionalities.
- **The number of process integrations successfully implemented.** The CoE monitors the number of successful process integrations to assess the seamless integration of CRM solutions with other sub-processes.
- **The rate of silo eliminations in CRM processes**. The CoE measures the rate silos are eliminated within CRM processes to ensure adequate information transfer and collaboration across the organization.

Function 2: Simplification[20]

This function is committed to the ongoing effort to reduce system complexities.

Principle	Best Practice	Standard
The organization has a formal Center of Excellence to manage governance and oversee delivery.	Determine the present role and scope of the customer relationship menagement CoE and empower it through executive sponsorship, funding, and people to oversee the CRM's governance and delivery of enhanced functionality.	Business and CRM teams decide together whether a change to the business process or CRM is better. Declarative solutions are fully considered before committing to a programmatic solution. Salesforce AppExchange products are fully considered by architects before building complex and difficult-to-maintain solutions.
The organization mitigates risks and removes redundancies.	Mitigate risks through review of administrative access to CRM, maintaining updated authentication functionality that prevents unauthorized access to systems, controlling and limiting user and developer access to each CRM instance and child instance and their functionality. These measures are based on functional responsibility, controlling how changes to the production system can be made and who can make them, making sure that all changes to production can be traced to the who, when, why, and how. Control measures also include establishing that not only are backup systems operational, but that functionality can quickly be restored through them. Control the number and access to sandboxes.	CoE facilitates efforts to reduce overlapping functions between CRM instances and between the CRM and other systems. CoE facilitates team agreements that create trust, autonomy, and purpose behaviors. CoE gears team processes and tooling to speed solution delivery.
We have a unified governance model and tool set across all our CRM platforms.	Establish a regular review of the organization's governance policies, standards, and procedures for CRMs, with accountability and responsibility assigned to communicating them across the CRM ecosystem. Due to proprietary platform architecture, CRM tool sets may be specialized and unusable in other IT systems and may not be usable by the rest of our IT systems. But within a single CRM platform, tool sets are standardized across CRM production instances.	CoE proposes work to decrease manual processes in CRM development and support. CRM teams stay current on Salesforce's latest functionalities and consider their impact on business solutions. Applications and tool sets are shared and reused across CRM production instances.

Figure 23: Example of the simplification function.

KPI examples for the simplification function:

- **Reduction in CRM system complexities**. The CoE measures the decrease in system complexities within the CRM environment.

- **Risk mitigation effectiveness in CRM governance.** The CoE assesses the effectiveness of risk mitigation strategies in CRM governance.

20   Alan Siegel and Irene Etzkorn, *Simple: Conquering the Crisis of Complexity* (New York: Hachette Books, 2013).
Peter Weill and Jeanne W. Ross, *IT Governance: How Top Performers Manage IT Decision Rights for Superior Results* (Boston: Harvard Business Review Press, 2004).
Al Decker and Donna Galer, *Enterprise Risk Management: Straight to the Point* (New York: Business Expert Press, 2013).

- **The number of redundancies removed in CRM processes**. The CoE tracks the number of CRM process redundancies eliminated.
- **Stakeholder satisfaction with CRM system simplification**. The CoE evaluates stakeholder satisfaction regarding the simplification of the CRM system.
- **The efficiency gained from a unified CRM governance model.** The CoE measures the efficiency improvements resulting from implementing a unified CRM governance model.

Function 3: Maintainability[21]

The function focuses on maintaining evolving systems without adding complexity or risk.

Principle	Best Practice	Standard
CRM solutions are evaluated for ease of maintenance and the required staff with skills to conduct ongoing modification.	Establish protocols for ease of maintenance and required skilled staff as criteria in evaluating solution options.	CRM solution design considers built-in universality, reuse, interchangeability, solution transparency, and documentation. Our people have the skill to readily make minor modifications to our existing CRM solutions as business needs change. Our maintenance people understand the built purpose for each of the CRM solutions they maintain.
Periodic CRM health checks are conducted and shared with relevant stakeholders.	Run continuous health assessments on CRM instances using the CRM's health check tools and third-party tools, and review and communicate with accountable parties at least monthly. Include defects discovered in health assessments in the enhancement road maps. Monitor CRM vendor communications for the newest standards or discovered issues and apply the recommended fixes. Keep the CRM code base updated to no more than two years older than the vendor's latest API version.	As our vendor updates and makes the CRM's default functionality easier to maintain, we rework our code to benefit from that. As our vendor updates and makes the CRM's default functionality more secure, we build that into our solutions. As our CRM vendor makes minor improvements to its customer experience, we incorporate the low-hanging improvements.
The tech debt in our organization is steadily being reduced.	Measure tech debt and establish goals with a timeline for reducing it. Include tech debt in the responsibilities of product management teams and allocate 20%–30% of team time to tech debt reduction.	Where CRM functionalities have overlapped and fewer CRM production instances were more efficient, we merged instances. Our CRMs have enough sandbox environments to manage development, testing, and prototyping. Before adding another CRM instance, the CoE, the business stakeholders, and the CRM vendor agree that it is required.

Figure 24: Example of the maintainability function.

KPI examples for the maintainability function:

- **Ease of CRM system maintenance (time and cost metrics).** The CoE measures the time and cost associated with maintaining the CRM system.
- **The success rate of periodic CRM health checks.** The CoE tracks the success rate of periodic health checks conducted on the CRM system.

---

21   Gene Kim, Kevin Behr, and George Spafford, *The Phoenix Project: A Novel About IT, DevOps, and Helping Your Business Win* (Portland, OR: IT Revolution Press, 2013).
Gene Kim, Jez Humble, Patrick Debois, and John Willis, *The DevOps Handbook: How to Create World-Class Agility, Reliability, & Security in Technology Organizations* (Portland, OR: IT Revolution Press, 2016).
Philippe Kruchten, Robert Nord, and Ipek Ozkaya, *Managing Technical Debt: Reducing Friction in Software Development* (Boston: Addison-Wesley Professional, 2019).

- **Reduction in technical debt over time.** The CoE monitors technical debt reduction within the CRM system over time.
- **The number of CRM modifications made in-house.** The CoE measures the number of CRM modifications successfully carried out by in-house teams.
- **System uptime and reliability metrics.** The CoE evaluates system uptime and reliability, ensuring the CRM system remains functional and dependable.

3. DOMAIN: FOUNDATION

This domain forms the core of the technical and strategic base for CRM activities.

Capability: Risk management

This capability is focused on limiting or removing risk to or from CRM systems through various functions.

Function 1: Risk mitigation[22]

This function limits or removes risk to or from CRM systems.

Principle	Best Practice	Standard
The organization has process controls in place for risk management.	Identify the kinds of risk to which a specific CRM system is exposed. Evaluate the organization's gaps in risk management process controls for each type of risk. Create a plan and a budget; delineate accountability and responsibility to mitigate each gap.	We maintain a project risk and opportunity log with a named owner for each risk mitigation item or opportunity play. Our planning documents and status reviews incorporate naming and updating project and business risks and opportunities. Our planning documents and status reviews include planned and updated responses to risks.
Risk management process controls are routinely tested in the organization.	Assign accountability and responsibility for routinely testing the organization's CRM risk-management process controls.	We plan for the emergence of new and reprioritization of existing business CRM requirements. CRM project stakeholders are regularly updated on risk status and efforts to mitigate risks. CRM teams regularly solicit updates from business on potential shifts that will change risk profiles.
The organization's risk management process controls are updated periodically.	Establish protocol, accountability, and responsibility for periodic updates to the organization's CRM risk-management process controls.	Tools like our CRM's health status are used to track reduction in platform technical risks. Corporate security experts are periodically consulted for risks that they see emerging for the platform. The CRM vendor is periodically solicited to review how implementation of their product might be at risk.

Figure 25: Example of the risk mitigation function.

KPI examples for the risk mitigation function:

- **The number of risks identified and mitigated**. The CoE tracks the number of identified and successfully mitigated risks.
- **Effectiveness of risk management controls (audit scores)**. The CoE measures the effectiveness of risk management controls through audit scores.
- **Frequency and success of risk management testing**. The CoE monitors how often risk management controls are tested and the success rate of these tests.

---

22   James Lam, *Enterprise Risk Management: From Incentives to Controls*, 2nd ed. (Hoboken, NJ: Wiley, 2014).
Kurt J. Engemann, *The Routledge Companion to Risk, Crisis and Security in Business* (London: Routledge, 2018).

- **Improvement in CRM risk profile over time**. The CoE evaluates the progress in the CRM risk profile, ensuring that risks are continuously reduced.
- **Stakeholder confidence in CRM risk management**. The CoE assesses stakeholder confidence in the organization's ability to manage CRM risks effectively

Function 2: Access control[23]

This function ensures appropriate CRM access for productivity while maintaining safety.

Principle	Best Practice	Standard
The organization's CRM implementation enforces role-based access and security, based on the "least privilege" principle.	Define and enforce role-based access and security by providing the least privilege required for the role.	Our end users are assigned access to the minimum amount of business data required for the task. Executives and managers are not presumptively assigned greater access. Their default is read-only access. We use sharing rules or other CRM system components to incrementally expand access to data for groupings of users, as the business case demands.
The segregation of duties in the security policy is well defined.	Define how duties are segregated and enforced within CRM security policies, standards, and procedures.	We conscientiously limit the number of system administrators and the places and kinds of changes they can make. Our system administrators log why they enter production and what they did there. Our people who are assigned system administrator roles are primarily focused on tasks requiring restricted access and have a reduced role in less-restricted tasks.
A transparent escalation process for violation of access control is defined and tested.	Define and test an escalation process for violations of access control.	Developers' and testers' access to functionality is specific to the environment and their role in that environment. Where a change to production requires a manual insertion, a system administrator does it—not a developer. A developer or tester with access to one environment in a CRM family doesn't, by default, have access to all such environments, unless their assigned tasks require this.

Figure 26: Example of the access control function.

KPI examples for the access control function:

- **The compliance rate with role-based access policies.** To ensure adherence to security protocols, the CoE monitors the compliance rate with role-based access policies.
- **Number of security breaches due to access-control violations.** To evaluate the effectiveness of current access controls, the CoE tracks the number of security breaches caused by access-control violations.
- **Effectiveness of segregating duties in CRM security.** To ensure proper access and reduce risk, the CoE assesses the effectiveness of segregating duties within CRM security.
- **Employee adherence to access-control policies.** The CoE measures employees' adherence to access-control policies to maintain security integrity.

---

23   Leighton Johnson, *Security Controls Evaluation, Testing, and Assessment Handbook* (Waltham, MA: Butterworth-Heinemann, 2015).

David Hillson, *The Risk Management Handbook: A Practical Guide to Managing the Multiple Dimensions of Risk* (London: Kogan Page, 2016).

- **Efficiency of escalation process for access violations.** To address and resolve security issues promptly, the CoE evaluates the efficiency of the escalation process for access violations.

Function 3: Business continuity[24]

This function focuses on building a resilient CRM infrastructure for business continuity.

Principle	Best Practice	Standard
The organization has business continuity plans that include resiliency, recovery, and contingency.	Define, write, and periodically update business continuity plans that include resiliency, recovery, and contingencies.	Business continuity CRM exercises consider security breaches, equipment failure, power outages, natural disasters, and missing staff. Business continuity CRM plan administrators collaborate with the CoE, business owners, security, and corporate continuity staff. CRM solutions and data are backed up regularly and incrementally, and test-restore exercises are conducted at least annually.
Restore exercises are conducted to include systems that interact with the CRM.	Design and conduct system, data, and function restoration exercises across the systems interacting with each CRM instance.	Restoration exercises are coordinated with teams from other systems that interact with the CRM. The architecture of restoration exercises is designed to include the restoration order of integrated systems and data sets. Testing and business confirm that restoration was successful.
A major incident plan is well-defined, with roles and processes.	Design, write, propagate, and keep updated a "major incident" plan with defined roles and processes.	CoE and business have plans for users to be able to perform some of their work should the CRM crash for a day or two. CoE and stakeholders have defined roles and plans for resolving emergent CRM issues that impact users or customers. Restoration planning includes our CRM and backup product vendors.

Figure 27: Example of the business continuity function.

KPI examples for the business continuity function:

- **Effectiveness of CRM business continuity plans.** The CoE measures the effectiveness of business continuity plans to ensure that resilience, recovery, and contingency strategies are robust and actionable.
- **The success rate of restoration exercises.** To validate the readiness and reliability of backup processes, the CoE tracks the success rate of system and data restoration exercises.
- **Stakeholder confidence in major incident plans.** To ensure that all involved parties understand and trust the roles and processes, the CoE assesses stakeholder confidence in major incident plans.
- **CRM uptime and availability metrics.** To maintain continuous operations and identify areas for improvement in system reliability, the CoE monitors CRM uptime and availability metrics.
- **Efficiency of emergency response and staffing coverage.** The CoE evaluates the efficiency of emergency response and staffing coverage to ensure that the organization can effectively manage and resolve emergent CRM issues.

---

24   Susan Snedaker, *Business Continuity and Disaster Recovery Planning for IT Professionals*, 2nd ed. (Waltham, MA: Syngress, 2013).

Capability: DevOps

This capability combines software development and IT operations for expedited delivery.

Function 1: Continuous development[25]

This function ensures frequent delivery of well-tested incremental changes to the CRM production environment.

Principle	Best Practice	Standard
The organization has an industry-standard methodology in place for continuous integration and continuous delivery and/or deployment (CI/CD)[26].	Create and deploy a CI/CD methodology and compatible CRM tools that provide for frequent delivery of new customer functionality by automating the stages of solution development. The CI/CD system should address continuous integration, delivery, and deployment.	We can deploy changes to production very frequently and are working to shorten that cycle time. Our new solutions can be reliably released at any time. Our updates are made in increments, enabling delivery as soon as they are complete and as soon as they pass automated predefined tests.
The organization delivers frequent incremental features securely.	Create a development process and deploy tools that securely deliver[frequent incremental features.	Continuous development automation is deployed in a manner to reduce risk and secure our solutions. Keys, credentials, and secrets are not included in scripts, source code, and text files. They are managed by a secret server, or alternate secure solution. Integration testing is a collaborative real-time process with other teams and is largely automated.
The organization has the capability to measure and improve on industry-standard DevOps metrics by how we handle broken solutions.	Create the capacity to measure the actual DevOps delivery metrics and to improve on them by how quickly and securely we can mitigate a broken solution that we have deployed.	We can identify all components and dependencies of a broken solution that needs to be backed out of production. Once broken solution components and dependencies are identified, our tools can automatically remove components from production. We have a well-defined approval process for removing broken solutions and testing that production works once again.

Figure 28: Example of the continuous development function.

KPI examples for the continuous development function:
- **Frequency and success rate of CI/CD.** The CoE measures how often and successfully instances of continuous integration and continuous delivery and/or deployment (CI/CD) occur to ensure that changes are reliably delivered to production.
- **Incremental feature delivery rate.** The CoE tracks the rate at which incremental features are delivered to production to maintain a steady flow of new functionalities.

---

25   Jez Humble and David Farley, *Continuous Delivery: Reliable Software Releases Through Build, Test, and Deployment Automation* (Boston: Addison-Wesley Professional, 2010).

26   Red Hat, "What is CI/CD?" December 23, 2023. https://www.redhat.com/en/topics/devops/what-is-ci-cd#:~:text=CI%2FCD%2C%20which%20stands%20for,a%20shared%20source%20code%20repository.

- **Adherence to industry-standard DevOps metrics.** The CoE assesses how closely its DevOps metrics align with industry standards to benchmark performance and identify areas for improvement.
- **Rate of automated testing implementation.** The CoE monitors the rate of automated testing implementation to ensure comprehensive coverage and reliability of CRM solutions.
- **Shorter time for development and deployment cycles.** The CoE evaluates the time to develop and deploy solutions to shorten cycle times and increase delivery efficiency.

Function 2: Testing and automation[27]

This function is about automated and manual CRM testing and other automation of development processes.

Principle	Best Practice	Standard
The organization has controls to avoid making unauthorized and unvetted changes directly in production.	Establish procedures, security access limits, and controls to avoid making changes directly in production. Define the exceptions, how they will be handled, by whom, and under what authorization.	We can deploy changes to production very frequently and are working to shorten that cycle time. Our new solutions can be reliably released at any time. Our updates are made in pieces, enabling delivery as soon as completed and as soon as they pass automated predefined tests.
The organization leverages native CRM platform automation.	Use native CRM platform automation to enhance user experience and adoption.	Continuous development automation is a part of reducing risk and securing our solutions. Keys, credentials, and secrets are not included in scripts, source code, and text files but are managed by a secret server. Integration testing is a collaborative real-time process with other teams and is largely automated.
The organization leverages the right tools to manage testing effectively.	Use CRM testing tools that effectively address declarative and programmatic code without requiring advanced developer skills to implement and maintain the system. Select testing tools that stay in lockstep with changes made by a vendor's CRM functionality, API versions, and nomenclature changes. This includes the ability to run regression tests with any code change in testing tool criteria.	Testing tools are regularly evaluated and updated to match the latest CRM platform version changes. Testing automation scripts are reviewed and optimized for efficiency and coverage. Training sessions are conducted regularly to ensure team members are proficient in using testing tools.

Figure 29: Example of the testing and automation function.

KPI examples for the testing and automation function:

- **The number of bugs the development team finds versus the end users**. The CoE tracks the ratio of bugs found by the development team compared to those found by end users to measure the effectiveness of prerelease testing.
- **Effectiveness of testing tools in CRM development**. The CoE assesses how well the selected tools support the development process and their ability to keep up with CRM platform changes.

27   Lisa Crispin and Janet Gregory, *Agile Testing: A Practical Guide for Testers and Agile Teams* (Boston: Addison-Wesley Professional, 2009).

- **Rate of reduction in direct production changes**. The CoE monitors the decrease in the number of direct production changes to ensure adherence to established procedures and controls.
- **Quality assurance metrics for CRM updates**. The CoE evaluates the quality of CRM updates through metrics for defect density, test coverage, and release stability.
- **Stakeholder satisfaction with CRM testing processes**. The CoE measures stakeholder satisfaction with the testing processes to ensure they meet expectations and contribute to reliable CRM performance.

Function 3: Talent development[28]

This function creates and evolves employee talent development programs to align with company needs.

Principle	Best Practice	Standard
The organization has defined CRM roles and responsibilities.	Define CRM roles and responsibilities and accountability for periodic review of those roles and responsibilities.	Our CoE has comprehensive job descriptions for each CRM role that clearly outline expectations, primary responsibilities, required qualifications, and performance metrics. It delineates junior, mid-level, and senior positions. Each CRM role is aligned with the organization's strategic objectives, and all stakeholders know the impact of each role on these objectives. We have a written responsibility assignment matrix, such as a RACI chart with these levels of involvement: responsible, accountable, consulted, informed. Our matrix is meant to clarify the roles and decision-making authority within CRM processes, eliminating ambiguity and fostering accountability.
The company offers CRM professionals a defined career path.	Define CRM career growth paths for each role with options for divergent paths.	We have a transparent career progression ladder that outlines the path from entry-level to senior leadership roles within the CRM domain, including the competencies and achievements required at each stage. We provide and fund continuous-learning opportunities, such as certifications, workshops, and courses, directly tied to the advancement criteria within the CRM career path. Clear performance metrics and promotion criteria directly link to career advancement, ensuring that promotions are based on merit and predefined standards of success.
The organization encourages its people to be professionally active outside the organization.	Create a plan for encouraging recognition with incentives for professional involvement by CRM talent outside the organization.	We encourage and support participation in CRM industry events, conferences, and seminars to foster professional networking and exposure to new ideas and practices. We recognize and reward employees who contribute to the broader CRM community through thought leadership activities, such as publishing articles, participating in speaking engagements, or active involvement in professional groups. Employees are incentivized to share knowledge and experiences from external activities with their internal teams, enriching the organization's CRM practices with fresh insights and perspectives.

Figure 30: Example of the talent development function.

KPI examples for the talent development function:

- **Alignment of CRM roles with organizational objectives.** The CoE ensures that all CRM roles and responsibilities align with the organization's strategic goals. The CoE also regularly reviews and updates

---

28   Ruth C. Clark, *Developing Technical Training: A Structured Approach for Developing Classroom and Computer-Based Instructional Materials* (Silver Spring, MD: International Society for Performance Improvement, 2012).

these roles to maintain alignment.

- **The success of CRM career growth paths.** The CoE measures the effectiveness of defined career paths by tracking employee promotions and advancements within the CRM domain.
- **Employee engagement in professional CRM activities.** The CoE monitors employee participation in external CRM industry events, conferences, seminars, and professional groups to judge professional engagement.
- **Impact of talent development on CRM initiatives.** The CoE evaluates how talent development programs contribute to the success and efficiency of CRM initiatives by assessing the performance and outcomes of these programs.
- **Rate of internal talent growth in CRM roles.** The CoE tracks internal talent growth within CRM roles by measuring the rate at which employees advance through the defined career path from entry-level to senior leadership positions.

Capability: Data architecture

This capability focuses on data-related design, management, and optimization.

Function 1: Data integration[29]

This function manages the effective integration of CRM with other systems.

Principle	Best Practice	Standard
The organization has a formal CRM data governance in place.	Create or enhance a CRM data governance plan that addresses privacy, accuracy, relevance, recovery, and access within and across systems.	Business is involved in mapping data flows and authorized access. Our data architecture requires masking of confidential data as it moves between systems and is made available to users. We have mapped the dependencies and the required order of data set restorations.
The CRM architecture is scalable and resilient.	Measure the CRM's performance speed and review for hiccups in the system that lock or slow down processing or will do so as the system expands. Remediate issues.	All integration points have non-person accounts and means of authentication. Confidential and legally restricted data is protected at each step in the data life cycle. Our solutions consider data integrity and performance velocity, impacted by record locking and sharing rules.
Data integration follows an established corporate data security policy.	Establish alignment with corporate data security policies when implementing CRM integrations. Review and update periodically. Stay up-to-date with CRM vendor's latest integration standards.	For mapping between records across systems, our data architecture does not depend on fields containing confidential values. We have tools for testing for data integrity, duplication, and restoration. All integrations have corollary test environments in the other systems, with appropriate masking of confidential data.

Figure 31: Example of the data integration function.

KPI examples for the data integration function:

- **Effectiveness of CRM data governance (compliance rate)**. The CoE ensures compliance with the formal CRM data governance plan by regularly monitoring and enforcing privacy, accuracy, relevance, recovery, and access standards across systems.
- **Scalability and resilience of CRM architecture.** The CoE evaluates CRM architecture's scalability and resilience by measuring performance speed, identifying system hiccups, and addressing issues that lock or slow down processing as the system expands.
- **Alignment with corporate data security policies**. The CoE establishes and maintains alignment with corporate data security policies for CRM integrations, regularly reviewing and updating practices to stay in line with the latest CRM vendor standards.

---

29   Anthony David Giordano, *Data Integration Blueprint and Modeling: Techniques for a Scalable and Sustainable Architecture* (Upper Saddle River, NJ: IBM Press, 2010).
John W. Foreman, *Data Smart: Using Data Science to Transform Information into Insight* (Hoboken, NJ: Wiley, 2013).

- **Data integration success rate across systems**. The CoE tracks the success rate of data integration efforts across systems by monitoring the seamless flow of data and the effective management of non-person accounts and authentication means at integration points.
- **Data flow efficiency and integrity metrics.** The CoE measures data flow efficiency and integrity by using tools to test data integrity, duplication, and restoration, ensuring that confidential and legally restricted data is protected at each step in the data life cycle.

Function 2: Data quality[30]

This function prevents and reduces duplicative data and maintains patterned cleanliness and usefulness of company data.

Principle	Best Practice	Standard
As the CRM platform releases greater data quality functionality, we implement those upgrades and better practices in our CRM systems.	Evaluate each functionality released by the CRM vendor for alignment with a CRM's business needs and implement relevant enhancements.	Data, security, integration, infrastructure, and mobile solutions have the flexibility to meet changing customer demands. We use our CRM vendor's health-check tools to ensure we meet more than minimum security requirements. As our CRM vendor publishes new best practice recommendations and functionalities, we review each for implementation.
We have tools to evaluate data quality with defined standards that must be met for code or data to move into production.	Conduct a workshop between developers and architects to determine the code standards to improve an organization's data quality. Implement a code quality scanning tool to enhance data security and quality before moving code into production.	Code quality automation tools are tuned to also evaluate code quality criteria that may impact data quality. Data quality tools are in place to catch duplicates and out-of-date information. Our code test coverage requirements include coverage for every use case.
CoE solution architects are embedded in all CRM solution teams, in-house or a vendor, and charged with including data quality in their solution evaluation.	Build into every CRM solution team, whether in-house or a vendor, an internal Center of Excellence architect responsible for data quality.	With criteria including how performance issues might impact data quality, solutions are vetted for performance issues before building, as solutions are built and as tested. Technical debt is planned for mitigation as we go, rather than waiting for a crisis to occur, and any technical debt that impacts data quality is prioritized. Mitigation of record locking, record-saving collisions, and other CRM system errors are built into solutions.

Figure 32: Example of the data quality function.

KPI examples for the data quality function:
- **Improvement in CRM data quality after enhancements.** The CoE measures the improvement in data quality after implementing enhancements by evaluating the cleanliness and usefulness of company data.
- **Effectiveness of data quality tools.** The CoE assesses the effectiveness of data-quality tools by monitoring their ability to catch duplicates, out-of-date information, and other data issues.

30   Jack E. Olson, *Data Quality: The Accuracy Dimension* (San Francisco: Morgan Kaufmann, 2003).

- **Alignment of data quality with CRM solution architecture.** The CoE ensures that data quality is aligned with the CRM solution architecture by embedding CoE solution architects in all CRM solution teams and making them responsible for data quality.

- **Rate of data quality issue resolution.** The CoE tracks the rate at which data quality issues are resolved, ensuring that technical debt and other system errors are mitigated as solutions are built and tested.

- **Stakeholder satisfaction with CRM data quality.** The CoE gauges stakeholder satisfaction with CRM data quality by conducting regular surveys and feedback sessions to ensure data meets business needs and security standards.

Function 3: Data standardization[31]

This function maintains standardized data naming, definitions, and usage across systems.

Principle	Best Practice	Standard
Implemented solutions have data documentation that can be referenced.	For each implemented solution, create useful documentation for those who will later maintain, enhance, or troubleshoot the solution.	All CRM build and support procedures, processes, content, and style guides are defined. We have an automation tool for periodically producing a data dictionary. As the solution is built, end user prompts are automated and built into the solution.
The CRM platform's built-in capacities for documenting the purpose and use of a field are well utilized.	Utilize the CRM platform's built-in capacities for documentation purposes and providing helpful information for end users. This should include definitions and descriptions of all data fields.	Declarative and programmatic code-naming conventions are defined by the CoE and used uniformly across all CRMs. Declarative fields and processes have description fields with helpful text, and the code is well commented on and traceable. Help fields have useful end-user information.
We have an architecture handbook with declared data and coding standards, best practices, and naming conventions.	Develop, maintain, and reference a data architecture handbook with declared data-naming standards, mastering practices, resolution logic for duplicate data, privacy, segregation, and aging-out rules.	All developers, whether in-house or a vendor, follow CoE coding standards and guides to avoid technical debt. CoE facilitates cross-function meetings to establish and document standards and working methods. All system integrators and contractors must work within our tool set, processes, and standards.

Figure 33: Example of the data standardization function.

KPI examples for the data standardization function:

- **Consistency in data documentation across CRM solutions.** The CoE ensures that all implemented solutions have consistent data documentation that can be referenced to assist with maintenance, enhancements, and troubleshooting.

---

31   Ralph Kimball and Margy Ross, *The Data Warehouse Toolkit: The Definitive Guide to Dimensional Modeling*, 3rd ed. (Indianapolis: Wiley, 2013).

Thomas H. Davenport and Jinho Kim, *Keeping Up with the Quants: Your Guide to Understanding and Using Analytics* (Boston: Harvard Business Review Press, 2013).)

- **Utilization rate of CRM's data field documentation capabilities.** The CoE maximizes the CRM platform's built-in capacities for documenting purposes and providing end user help information for all data fields.
- **Adherence to data and coding standards.** The CoE enforces uniform declarative and programmatic code-naming conventions across all CRMs, ensuring all developers follow CoE coding standards and guides to avoid technical debt.
- **Rate of data standardization improvements.** The CoE measures improvements in data standardization by tracking the implementation and maintenance of declared data naming standards, mastering practices, undoing-duplication logic, privacy and segregation rules, and age-out rules.
- **Efficient utilization of data architecture handbook.** The CoE develops, maintains, and regularly references a data architecture handbook with declared data and coding standards, best practices, and naming conventions, ensuring that all developers and system integrators comply.

Conclusion: A foundation for implementation

Chapter 2 laid a robust foundation for implementing the CoE framework with first principles, best practices, standards, and KPIs for specific functions within the CoE realm of interest.

Focusing on a select number of functions at any given time ensures that enhancement efforts are manageable and effective. This approach helps organizations streamline their efforts, align with corporate policies, and achieve sustainable success.

The next chapter will introduce a structured approach to determining which of the 27 functions most significantly influence business outcomes. Employing this framework to periodically measure and track progress within the CRM CoE realm will iteratively highlight the next functions to prioritize for improvement. Accelerating focused decision-making in this context is invaluable. It propels organizations toward the next iteration of goals for enhancing business outcomes.

(With attribution to book, author, and copyright, these cookbook section recipes for functions may be used. They are available as images at www.crmcoe.com/book-bonuses.)

CHAPTER 3
Iterative Transformation

Executive summary

Chapter 3 offers a structured methodology for the iterative transformation of a CRM Center of Excellence (CoE). It outlines the steps to refine and elevate customer relationship management (CRM), efficiently prioritizing the CoE's next set of functions that will most impact business outcomes.

The framework involves assessing the CoE's current state and prioritizing functions for improvement. As these functions are improved, periodic evaluations identify the next set of functions to improve the system, creating a continuous improvement cycle. The framework also addresses weaknesses within the CoE's cultural aspects—data, people, process, technology, and agility. Addressing sets of functions through this lens builds a more resilient organization.

Maturing a function involves implementing best practices aligned with core principles, supported by standards and tracked by key performance indicators (KPIs) for validating improvements and making strategic adjustments. The business outcome ladder visually represents the sequence of this improvement process, focusing, in order, on reducing risk, improving user adoption, lowering operational expenses, increasing the Net Promoter Score (NPS), and growing revenue. This structured approach ensures foundational improvements first, creating a robust base for more complex outcomes.

By systematically applying the principles, best practices, standards, and KPIs outlined in this chapter, organizations can ensure continuous improvement and alignment with evolving business objectives, ultimately driving efficiency and effectiveness across all functions. This methodical enhancement of CRM CoE functions promotes sustainable progress and significant improvements in business outcomes.

Business outcomes

We visualize improving business outcomes within a CRM CoE as ascending a Business Outcomes Ladder with five rungs, each representing a core business outcome: reducing risk, improving user adoption, lowering operational expenses, increasing the NPS, and growing revenue. The journey to enhance these outcomes begins by identifying and fortifying the weakest functions at the lowest rung of this metaphorical ladder. This approach ensures that foundational improvements are secured first, creating a robust base for more complex and higher-level outcomes.

The initial step in this climb is the assessment process, pinpointing the functions that directly impact the foundational business outcomes. By concentrating efforts on these weak points, organizations can achieve immediate, tangible improvements, effectively setting the stage for tackling subsequent challenges farther up the ladder.

Functions driving business outcomes

Functions impacting business outcomes are sequential, building up the Business Outcomes Ladder but not necessarily influencing outcomes downward. Functions like data integration, data quality, data standardization, business continuity, access control, and risk mitigation drive the business outcome of reduced risk. Enhancing one of these functions can likely lead to improvements in the other functions due to their synergistic effects.

Prioritizing functions for impactful business outcomes

Prioritizing CRM CoE functions is essential to align the Center of Excellence's operations with an organization's specific needs and objectives. This process starts with a thorough assessment to identify areas significantly impacting foundational business outcomes, enabling targeted and impactful enhancements.

Addressing all twenty-seven functions simultaneously is neither feasible nor practical. Instead, choosing a subset aligned with any immediate challenges and priorities ensures that improvement efforts are manageable and directly beneficial to an organization's goals, and this promotes sustainable progress.

Assessment and bottom-up improvement

Using this framework, the CRM CoE assessment aims to identify key priorities and focus areas by analyzing team feedback and pinpointing critical functions that need improvement. This foundational step is crucial for determining areas that need attention. Addressing lower-level business outcomes, such as reducing risk, impacts operational expenses and can enhance higher-level outcomes like revenue growth by streamlining processes and reallocating resources towards growth initiatives. Focusing on a top-rung business outcome like revenue growth before managing bottom-rung business outcomes like reducing risk can produce immediate but unsustainable revenue growth.

Measuring business outcomes

Measuring the impact on business outcomes is vital for justifying the CoE's business value and guiding its strategic direction. This involves validating best practices, developing and measuring KPIs for targeted functions, and continuously refining KPI strategies based on feedback and results. If KPIs for functions go in one direction and the business outcomes metrics go in the opposite direction, it's crucial to examine what's being measured and why. Are the KPIs measuring the correct factors or are the ideals that are being implemented not aligned to the desired business outcomes?

CoE Cultural Aspects: Fostering identity and balance

The Cultural Aspects of a Center of Excellence encapsulate its identity, emphasizing the importance of maintaining balance across five areas: data, people, process, technology, and agility. This holistic approach ensures that the CoE remains agile, aligned with business outcomes, and primed for adaptation to meet future demands, thereby nurturing sustained growth and evolution across the CoE.

Maturing the CRM CoE through a structured approach

The journey towards maturity of a CRM CoE is intentional and planned, rooted in the Five-Step Process for CRM CoE refinement. This process begins with selecting functions for refinement, drilling down into the chosen functions, defining ideals, executing improvements, and validating and iterating.

Culminating in organizational transformation

Employing the Five-Step Process for CRM CoE refinement, from raising awareness about the CoE's role to encouraging a culture of perpetual progress, the CoE substantially elevates its influence on business outcomes. Adhering to this flexible, well-considered strategy and sustaining a cycle of assessment, enhancement, and communication, the CoE is well-positioned to catalyze positive organizational transformation and harmonize its endeavors with broader enterprise goals.

Prioritizing CRM CoE functions

The CRM CoE Framework provides a structured methodology for advancing the CoE's maturity process. It begins with clearly recognizing the CoE's essential role within the organization and progresses through strategic steps designed to refine and elevate the CoE's value realization.

At the heart of this approach is prioritizing CRM CoE functions that will most impact business outcomes. By focusing on these critical areas, the CoE can efficiently channel its improvement efforts to achieve the highest returns.

In tandem with this focus, the framework also addresses areas of weakness within the CoE's Cultural Aspects, aiming to strengthen these facets across functions to build a more resilient organization. The CRM CoE's Cultural Aspects consist of five subsets of functions that span the CoE realm of interest, providing a holistic view of the CoE's landscape.

Maturing a function involves assessing its current state and prioritizing specific components for enhancement. This process includes implementing best practices aligned with cardinal principles for each function, supported by standards that uphold these practices, and tracked by key performance indicators (KPIs) for each practice. These KPIs enable continuous monitoring and strategic adjustments to align with evolving business objectives, ensuring that the CoE improves and adapts to changing demands and opportunities.

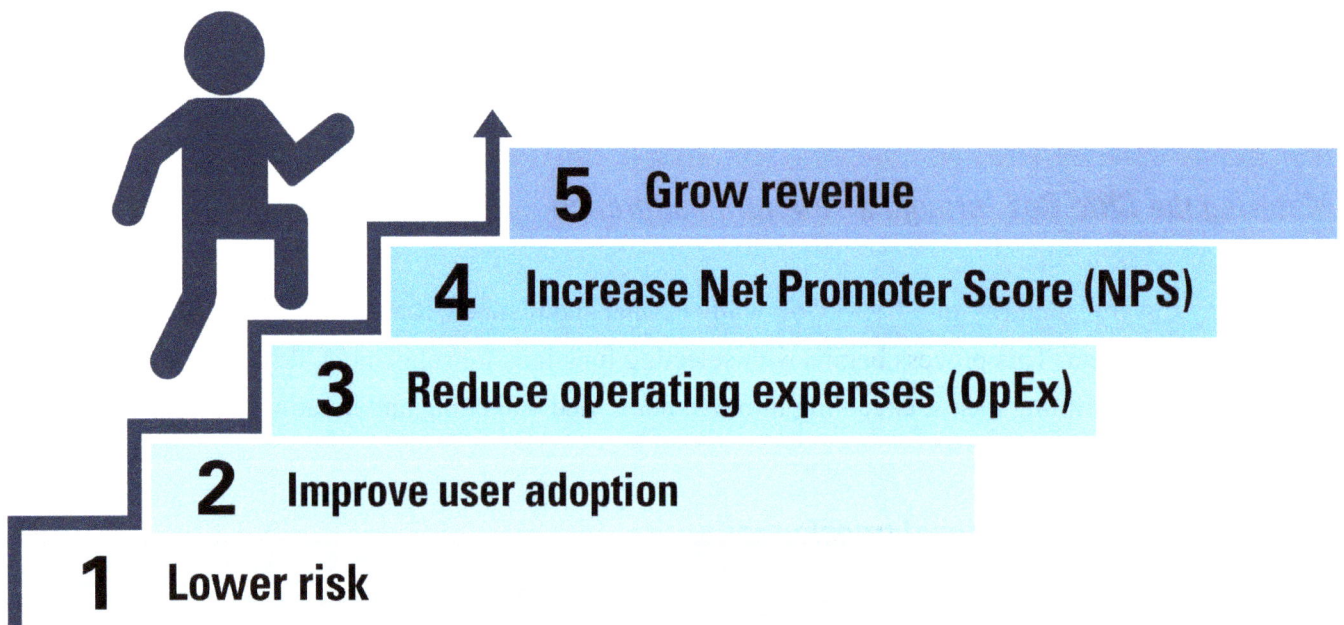

Figure 34: Business outcomes ladder, from the bottom up.

Improving business outcomes within a Center of Excellence can be visualized as ascending a ladder with five rungs, each representing a core business outcome: lowering risk, improving user adoption, reducing operational expenses, increasing the Net Promoter Score, and growing revenue. The journey to enhance these outcomes begins by identifying and fortifying the weakest functions at the lowest rung of this metaphorical ladder. This approach ensures that foundational improvements are secured first, creating a robust base for more complex and higher-level outcomes.

The initial step in this climb is the assessment process, pinpointing the functions that directly impact the foundational business outcomes. By concentrating efforts on these weak points, organizations can achieve immediate, tangible improvements, effectively setting the stage for tackling subsequent challenges represented by the higher rungs of the ladder. For instance, reducing customer acquisition costs is a crucial function that falls under the broader business outcome goal of reducing operating expenses (OpEx). Organizations can lower customer acquisition costs by optimizing marketing strategies and improving lead conversion rates, enhancing overall cost efficiency. Achieving this may require not only work on the data quality function but also on data relevance to the marketing function. So if we are working on data quality and we know that customer acquistion cost is something that we wish to impact, we would also look at the relevance of that data to marketing and perhaps to sales.

As organizations embark on this upward journey, they encounter challenges, such as limited resources and resistance to change. Addressing these requires a cross-functional team built around clear communication, collaborative work, and strategic alignment, akin to establishing a team of mountain climbers in sync with each other. Their synergy motivates them to take turns securing ropes above for those who follow, making the ascent safely, supporting each other in overcoming obstacles, and successfully reaching the top. Synergistic cross-functional teams act to provide each other the ropes and harnesses necessary to ensure no function is left behind.

The strategic focus on the weakest functions at the lowest rung, such as enhancing data quality or improving user adoption, lays a solid foundation for achieving the next level of business outcomes. For example, better user adoption (a foundational improvement) leads to more accurate data capture and more effective customer interactions, which, in turn, can enhance customer satisfaction (NPS) and ultimately contribute to revenue growth. This sequential improvement underscores the interconnectedness of the rungs and the importance of starting from the bottom up.

Navigating this climb involves addressing immediate functional weaknesses, continuously measuring progress through KPIs, and adapting strategies based on these insights. The ongoing assessment and adjustment ensure that each step taken up the ladder is secure and contributes to the overarching goal of enhancing business outcomes. However, the competence of the individual CRM CoE function and the integrated work of functions across CoE Cultural Aspects—data, people, process, technology, and agility—play a pivotal role in a sustained ascent.

Functions driving business outcomes

Functions impacting business outcomes are sequential, building as we go up the Business Outcomes Ladder but not necessarily influencing outcomes below. The data quality function, for example, affects all five business outcomes, starting with reducing risk and extending up to support revenue growth. While the marketing function, directly influences revenue growth, a business outcome above it, it is not likely to impact data quality, a business outcome below it.

Functions such as data integration, data quality, data standardization, business continuity, access control, and risk mitigation drive the business outcome of reduced risk. These functions have a synergistic effect: Enhancing one can likely lead to improvements in the others. Some of these synergies are more evident than others; for instance, achieving advancements in data integration and standardization typically yields improvements in data quality.

Domain	Capability	Function	Reduce Risk	Increase User Adoption	Lower Operating Costs	Increase Net Promoter Score	Grow Revenue
Customer Centricity	Vision & Strategy	Goal setting & prioritizing				X	X
		Transparency				X	X
		Purpose-driven				X	X
	Go-to-market	Marketing					X
		Sales operations					X
		Service center					X
	Innovation	Ideation				X	X
		Experimentation				X	X
		Value creation				X	X
Operations	PMO	Talent management			X	X	X
		Budget			X	X	X
		Product road map			X	X	X
	Change Management	Change control			X	X	X
		Adoption			X	X	X
		Agile framework			X	X	X
	Alignment	Workflow/Process map		X	X	X	X
		Simplification		X	X	X	X
		Maintainability		X	X	X	X
Foundation	Risk Management	Risk mitigation	X	X	X	X	X
		Access control	X	X	X	X	X
		Business continuity	X	X	X	X	X
	DevOps	Continuous development		X	X	X	X
		Testing & automation		X	X	X	X
		Talent development		X	X	X	X
	Data Architecture	Integration	X	X	X	X	X
		Data quality	X	X	X	X	X
		Data standardization	X	X	X	X	X

Figure 35: Business outcomes impacted by functions.

Case studies of the connection between functions and business outcomes

Business outcomes such as reducing risk, improving user adoption, lowering operational expenses, increasing the NPS, and growing revenue are benchmarks for the CoE's impact and directing its efforts toward the organization's strategic goals. Below are several cases illustrating the relationship between functions and business outcomes.

Case #1. Improved data quality and higher customer satisfaction

A retail company's decision to enhance its CRM system's data quality through stringent data validation processes is not just a technical improvement. It's a strategic move that leads to more accurate customer profiles and personalized marketing campaigns, directly increasing customer satisfaction and loyalty. This, in turn, results in a significant boost in the company's NPS and repeat purchase rates, demonstrating the direct impact of improved functions on business outcomes.

Case #2. Enhanced user adoption and reduced operational expenses

A financial services firm's investment in comprehensive training programs and user-friendly CRM interfaces is not just a cost. It's a strategic decision that leads to improved user adoption rates. Employees find the system easier to use and more helpful in daily tasks, so the firm experiences reduced operational expenses. This is due to decreased time spent on manual data entry and fewer errors in customer data management, leading to more efficient service delivery.

Case #3. Proactive risk management and increased revenue

A health-care provider implements a proactive risk management strategy within its CRM system, focusing on compliance with patient data regulations. By regularly auditing and updating their data security measures, they reduce the risk of data breaches and regulatory fines. This protects the company from financial losses and builds trust with patients and partners, increasing patient retention and attracting new patients. The enhanced reputation and trust translate into higher revenue growth for the health-care provider.

Case #4. Optimized workflows of an adviser and increased client satisfaction

An investment adviser needed multiple screens, which provided the contextual information he needed to advise clients. This cumbersome process led to delays and frustration. Switching to a multisource dashboard within the CRM streamlined the process and reduced the adviser's time per client, which allowed the adviser to address more clients per day. The result was higher operational efficiency and, thus, more client satisfaction, directly impacting business outcomes.

Case #5. Intelligent claims routing and enhanced processing efficiency

In a health-insurance company, due to the complexity of reimbursement rules and the failure to match the medical knowledge of the claims adjuster to the claim type, it was difficult for claims adjusters to identify the rationale behind claim rejection or reductions. By building a rule-based routing system within the CRM to match claims with available, knowledgeable adjusters, the company improved accuracy in claim processing and reduced time spent on duplicative work. This led to meeting contractual turnaround times more consistently, processing more claims per day with less staff, and higher provider and customer satisfaction ratings. This improvement showcased the impact of reducing operational expenses and increasing customer satisfaction. Note that today, if we were building or rebuilding this rules-based routing system, we would want to incorporate AI in administering the rules and helping the claims adjusters do their job by highlighting the pertinent decision-making information in the claim.

Summary

The journey to a mature CRM Center of Excellence is structured and deliberate, aimed at aligning the CoE's initiatives with the organization's unique needs and strategic goals. This journey begins with a comprehensive assessment to identify the critical functions impacting foundational business outcomes, setting the stage for targeted improvements. By focusing on functions that offer significant leverage on these foundational outcomes, we pave the path for systematically addressing broader organizational objectives.

CoE Cultural Aspects: Fostering identity and balance

The People, Process, Technology (PPT) framework, developed in the 1960s by business management expert Harold Leavitt and popularized in the late 1990s by information security technologist Bruce Schneier, has evolved into a staple concept of project management and change management thinking. Inspired by this framework, we have placed data at the forefront of the customer relations CoE framework and introduced agility. Thus, this refined model identifies data, people, process, technology, and agility as the five Cultural Aspects of a holistic approach to CoE maturation.

The configuration and weighting the CoE gives the Cultural Aspects encapsulate the center's identity. Although organizations will endow each of the aspects with greater or lesser importance and prioritize them differently, it is important to maintain balance across the five cultural aspects. Unlike the linear alignment of domains or capabilities, each cultural aspect draws on relevant functions across the entire realm of interest. This approach underlines the crucial roles of human elements, operational efficiency, technological advancement, data integrity, and adaptive innovation in defining the CoE's culture.

CRM CoE Cultural Aspects

Data	People	Process	Technology	Agility
Integration	Purpose-driven	Marketing	Budget	Goal setting & prioritizing
Data quality	Sales operations	Service center	Change control	Transparency
Data standardization	Ideation	Product road map	Risk mitigation	Experimentation
	Talent management	Workflow/Process map	Business continuity	Value creation
	Adoption	Simplification	Testing & automation	Agile framework
	Maintainability			Continuous development
	Access control			
	Talent development			

Figure 36: Functions organized in buckets according to Cultural Aspects.

CRM CoE Cultural Aspects: Pulling from across domains and capabilities

Data: As the cornerstone of decision-making, data underscores the importance of high-quality, well-documented, standardized, and accessible information.

People: The organization's most valuable asset is developing and engaging personnel and fostering a culture that supports continuous improvement and innovation.

Process: Efficiency in processes ensures streamlined operations and quality outcomes.

Technology: A pivotal enabler for business growth and operational capability.

Agility: Represents the capacity for rapid adaptation, fostering a culture of innovation and responsiveness.

CRM CoE Cultural Aspects Template

Domain	Capability	Function	Data	People	Process	Technology	Agility
Customer Centricity	Vision & Strategy	Goal setting & prioritizing					✓
		Transparency					✓
		Purpose-driven	✓				
	Go-to-market	Marketing		✓			
		Sales operations	✓				
		Service center			✓		
	Innovation	Ideation	✓				
		Experimentation					✓
		Value creation					✓
Operations	PMO	Talent management		✓			
		Budget				✓	
		Product road map			✓		
	Change Management	Change control				✓	
		Adoption	✓				
		Agile framework					✓
	Alignment	Workflow/Process map			✓		
		Simplification			✓		
		Maintainability	✓				
Foundation	Risk Management	Risk mitigation				✓	
		Access control	✓				
		Business continuity				✓	
	DevOps	Continuous development					✓
		Testing & automation				✓	
		Talent development	✓				
	Data Architecture	Integration	✓				
		Data quality	✓				
		Data standardization	✓				

Figure 37: Cultural aspect buckets across capabilities and domains.

The framework's focus on Cultural Aspects is a strategic method for sustaining long-term achievements by integrating and balancing these aspects across the CoE. Continual assessment to determine the weakest aspect is essential for holistic improvement, even as you focus on improving specific functions based on business outcome criteria.

Improving functions, applied through the lens of Cultural Aspects

The following case studies illustrate the success of this practice, aspect by aspect.

Case #6. Improved data quality and higher customer satisfaction

A retail company's decision to enhance its CRM system's data quality leads to more accurate customer profiles and personalized marketing campaigns, directly increasing customer satisfaction and loyalty. This, in turn, boosts the company's NPS and repeat purchase rates.

People aspect: By involving employees in data validation and emphasizing the importance of accurate data entry, the organization can foster a culture of accountability and precision. Regular training and recognition programs for employees who excel in maintaining data quality can further enhance engagement and productivity, leading to sustained upticks in customer satisfaction.

Case #7. Enhanced user adoption and reduced operational expenses

A financial services firm's investment in comprehensive training programs and user-friendly CRM interfaces leads to higher user adoption rates and reduced operational expenses.

Process aspect: Streamlining and integrating the training process with day-to-day operations ensures that employees can learn and apply new skills efficiently. Continuous feedback loops and process refinements based on user experiences can reduce training times and operational errors, further lowering expenses and enhancing service delivery.

Case #8. Proactive risk management and increased revenue

A health-care provider implements a proactive risk management strategy within its CRM system, focusing on compliance with patient data regulations. This reduces the risk of data breaches and regulatory fines, builds trust, increases patient retention, and attracts new patients.

Technology aspect: Leveraging advanced CRM technologies for real-time risk assessment and compliance monitoring can enhance the provider's ability to manage patient data securely. Continuous technological upgrades and integration with other health-care systems can ensure the CRM remains robust and compliant, driving higher revenue growth through enhanced trust and efficiency.

Case #8. Optimized adviser workflows and increased client satisfaction

An investment adviser who previously struggled with multiple screens to gather information now benefits from a multisource dashboard within the CRM. This reduces time per client, allowing advisers to handle more clients and improve satisfaction.

Data aspect: Ensuring the CRM dashboard provides accurate, real-time, valuable data from multiple sources enhances the adviser's ability to make informed decisions quickly. Standardizing data sources and improving data integration can provide more reliable information, increasing adviser efficiency and client satisfaction.

Case #9. Intelligent claims routing and enhanced processing efficiency

A health-insurance company improves claims processing accuracy and reduces the amount of work that must be redone by routing claims to knowledgeable adjusters through a rule-based system within the CRM.

Agility aspect: Building an adaptive system that can quickly respond to changes in claim types or adjuster availability ensures the CRM remains effective under varying conditions. Agile methodologies in system development and regular updates based on user feedback can maintain high efficiency and adaptability, improving processing times and customer satisfaction.

Summary

This holistic view, where cultural aspect criteria influence efforts that span various capabilities and domains, facilitates effective decision-making and balanced strategic CRM practices. As a result, business outcomes become more sustainable and resilient. Focusing on these Cultural Aspects ensures cohesion, sustainability, and adaptability, significantly contributing to the long-term success and impact of the CoE. Organizations can drive continuous improvement by addressing the weakest cultural aspect, leading to better business outcomes and a robust, adaptable CRM ecosystem.

Figure 38: The Five-Step Process iterating CRM CoE refinement.

Improving the CRM CoE with a Five-Step Process

The journey toward maturing a CRM Center of Excellence is intentional and planned. The objective is to synchronize the CoE's activities with the organization's distinct needs and strategic aspirations. This process, rooted in the Five-Step Process for CRM CoE refinement, as mentioned in Chapter 1, is the backbone of the CoE's development road map.

These are the steps in the process:

Step 1: Select functions for refinement.

The path to maturity begins with a thorough assessment identifying which of the twenty-seven functions are pivotal in affecting foundational business outcomes. Such an evaluation sets the trajectory for enhancements that will systematically unfold to address broader, more strategic goals. This initial step ensures that only a subset of functions with the most significant promise for immediate and impactful change is chosen for improvement, making the task realistic and strategically aligned.

Step 2: Drill down into the selected functions.

Once a subset of functions is identified, a deep dive is undertaken to understand their current state and pinpoint areas ripe for development. This process scrutinizes existing performance and uncovers opportunities for improvement, ensuring that the initiative's scope is targeted.

Step 3: Define ideals.

Moving forward, a structured framework encompassing principles, best practices, standards, and KPIs is established for the prioritized functions. This framework guides the enhancement strategy and is referenced throughout the implementation phase.

Step 4: Execute improvements.

A comprehensive strategy for fortifying the chosen functions is then articulated, underpinned by robust justifications focusing on tangible business outcomes. This plan is implemented and promoted throughout the organization to ensure buy-in and support.

Step 5: Validate and iterate.

Finally, a cycle of evaluation and refinement begins, marked by regular reviews of the effectiveness of the enhancement strategies against KPIs and evolving business goals. This iterative process can be adjusted to fit each organization, ensuring the CoE's actions are in lockstep with their organization's overarching objectives and that the CoE's role continues to create benefits.

In the context of the CoE's development, it is advisable to pay attention to the balance of the CoE's Cultural Aspects. This equilibrium is critical in ensuring the organization is ready to navigate an ever-evolving landscape where we don't know the direction for the next challenge. High-level assessments using the business outcomes perspective are the compass for achieving continuous and targeted improvement efforts.

Delusion testing as part of validation

In large companies, business case creators and approvers may differ from the delivery teams, often including systems integrators. Interpreting the original business case and project charter can be challenging. Due to a combination of factors, from time constraints and budgetary concerns to leadership changes, competence of the delivery team, and a failure to fully understand the initial vision, implementation of the original business case requirements is frequently less than stellar. Our experience in CRM implementations reveals a significant variance between the original business case and the final implementation. Therefore, a method for testing the process is necessary to optimize investments and ensure alignment with corporate objectives. Delusion testing is one such realty check.

Delusion Testing

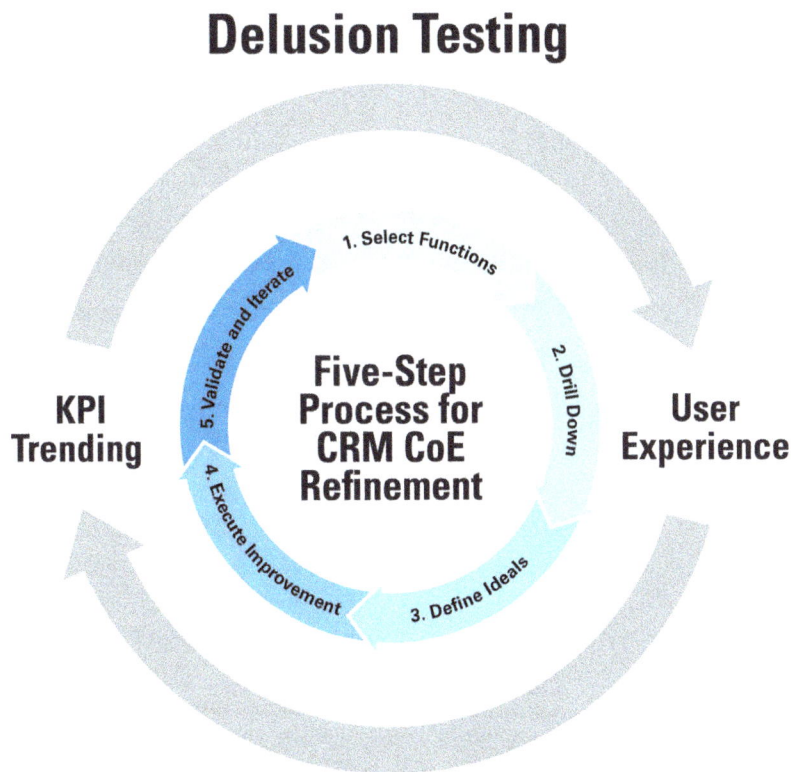

Figure 39: KPI delusion testing: Do positive KPI trends represent stakeholder experience?

Delusion testing is a vital process in the Center of Excellence that's designed to ensure that the perceived success of CRM initiatives, as indicated by KPI trend lines, truly aligns with users' experience across the organization. This involves:

1. **Initial and periodic assessments.** The process begins with assessments conducted by cross-functional team members to gauge the CRM CoE's effectiveness from a broad range of perspectives. These evaluations help pinpoint and prioritize weak functions impacting business outcomes.

2. **Real-world user experience verification.** At the heart of delusion testing is our commitment to ensuring that positive KPI trends reflect improved user experiences. For instance, while KPIs might indicate accurate, duplicate-free, and well-integrated data, an investment adviser or claims agent might not find this data timely, relevant, or useful for their needs. This discrepancy prompts us to reevaluate, ensuring that our KPIs measure what matters most to the users and that our standards and best practices align with their needs and operational styles.

3. **Iterative refinement process.** Our testing is not a one-time event but a continuous improvement journey. We refine the functions of the CRM CoE based on ongoing assessments and real-world feedback, always striving to align the CRM CoE realm of experience with what the KPIs measure. This commitment to continuous improvement instills confidence in our ability to deliver business value effectively.

Delusion testing challenges the assumptions underlying CRM performance metrics to prevent complacency. This rigorous scrutiny ensures that improvements in the CRM CoE functions provide real user benefits and, ultimately, business value.

This process culminates in organizational transformation employing the Five-Step Process, from raising awareness of the CoE's role to endorsing a culture of perpetual progress. Ultimately, the CoE substantially elevates its influence on business outcomes. Adhering to this flexible, well-considered strategy and sustaining a cycle of assessment, enhancement, and communication, the CoE is well-positioned to catalyze positive organizational transformation and harmonize its endeavors with the larger goals of the enterprise.

Conclusion

Chapter 3 presents a comprehensive iterative approach to transforming a CRM Center of Excellence through a structured and strategic methodology. It underscores the necessity of continuous assessment and iterative refinement, ensuring that CRM initiatives are always aligned with strategic goals. This focus on continuous improvement is vital to creating a resilient and agile CoE.

The framework detailed in this chapter emphasizes the Business Outcomes Ladder, a crucial tool that guides the sequential improvement from foundational outcomes to higher-level objectives. Organizations can ensure that their CRM initiatives are aligned with their strategic goals by addressing areas of weakness within the CoE's Cultural Aspects and employing best practices supported by robust standards and KPIs. Ultimately, by adhering to the flexible strategy of the Five-Step Process for CRM CoE refinement and sustaining a cycle of assessment, enhancement, and communication, the CoE is well-positioned to catalyze positive organizational transformation.

CHAPTER 4
Value Realization with the CRM CoE

Executive Summary

Chapter 4 explores economic strategies crucial for funding and improving CRM Center of Excellence (CoE) functions. These strategies—cost avoidance, savings, reduction, and optimization—ensure financial health and drive significant business benefits.

Financial analysis is crucial to the success of the Business Outcomes Ladder strategy, enabling CRM CoE leaders to justify funding requests by showing how specific enhancements contribute to strategic business outcomes and optimize costs. Beyond cost management, we explore how identified CRM CoE functions can improve Net Promoter Scores and drive top-line revenue growth, transforming CRM CoEs from acting as cost centers to acting as strategic top-line assets.

This chapter examines CRM CoE maturation through a cost lens. Then, it considers how investments can be focused on obtaining measurable contributions to customer loyalty and revenue growth.

Financial strategies for cost avoidance, savings, reduction, and optimization are crucial for increasing long-term net operating income. While related, each of these strategizes has unique implications for specific functions within the CoE framework. It is essential to also explore opportunities that directly contribute to top-line revenue growth, ensuring a balanced approach to financial health and value realization[32].

Cost avoidance focuses on preempting potential costs, preventing unnecessary expenditures before they occur.

Cost savings typically involve actions that result in reduced expenditure without compromising the quality or quantity of outcomes, akin to finding a more efficient route up the ladder without sacrificing safety or the goal of reaching the top.

Cost reduction, conversely, entails cutting expenses by scaling back the operations' scope, quality, or scale. This would be like choosing a less ambitious path up the ladder, skipping over rungs that seem less crucial but could significantly impact long-term success.

Cost optimization involves analyzing and adjusting spending to maximize value, ensuring resources are allocated wisely.

Incorporating this financial analysis into the Business Outcomes Ladder strategy adds weight to funding requests for improving targeted functions. By demonstrating how specific enhancements at each rung of the ladder contribute to both achieving strategic business outcomes and optimizing costs, CRM CoE leaders present a compelling investment case. For instance, enhancing user adoption (a lower rung on the ladder) through training and support can lead to better utilization of the CRM system and leverage of CRM investment. This can reduce the need for additional software purchases or custom developments to compensate for underutilization. Such discussions underscore the financial prudence of fully investing in CRM CoE improvements and their adoption by users.

Leaders can more effectively justify funding requests by illustrating how enhancements to targeted functions contribute to long-term increased net operating income through cost optimization. Such illustrations

32   Jack Freund and Jack Jones, *Measuring and Managing Information Risk: A FAIR Approach* (Waltham, MA: Butterworth-Heinemann, 2014).
William D. Pace, *Return on Investment (ROI): Basics for Consultants* (Boca Raton, FL: CRC Press, 2009).
Thijs Homan, *Organization Dynamics in the Network Era* (London: Springer, 2017).
Douglas W. Hubbard, *How to Measure Anything: Finding the Value of Intangibles in Business*, 3rd ed. (Hoboken, NJ: Wiley, 2014).
Gene Kim, *The Unicorn Project: A Novel about Developers, Digital Disruption, and Thriving in the Age of Data* (Portland, OR: IT Revolution Press, 2019).

demonstrate the CoE leadership's grasp of the strategic alignment between CoE initiatives and broader organizational financial goals. It frames the case for investment, not merely as a request for funds but as financial management aimed at extracting maximum value from CRM initiatives.

Domain	Capability	Function	Financial Justification Strategies				
			Cost Avoidance	Cost Savings	Cost Reduction	Cost Optimization	Revenue Growth
Customer Centricity	Vision & Strategy	Goal setting & prioritizing	X				
		Transparency		X			
		Purpose-driven	X				
	Go-to-market	Marketing					X
		Sales operations					X
		Service center			X		
	Innovation	Ideation					X
		Experimentation					X
		Value creation				X	X
Operations	PMO	Talent management		X			
		Budget		X			
		Product road map					X
	Change Management	Change control	X		X		
		Adoption					X
		Agile framework	X			X	
	Alignment	Workflow/Process map	X			X	
		Simplification		X	X		
		Maintainability		X			
Foundation	Risk Management	Risk mitigation	X	X	X		
		Access control	X	X			
		Business continuity	X	X			
	DevOps	Continuous development				X	
		Testing & automation		X		X	
		Talent development				X	X
	Data Architecture	Integration	X			X	
		Data quality	X			X	
		Data standardization	X				

Figure 40: Financial justification strategies per function.

Cost avoidance

Certain CRM CoE functions are well suited to a cost-avoidance strategy where proactive measures can prevent unnecessary future expenditures.

Ensuring high **data quality** and consistency across systems helps prevent costly data issues that could lead to poor decision-making, customer dissatisfaction, or compliance failures. Proactively managing data quality avoids expenses associated with rectifying errors, redundant data cleanup, additional compliance checks, and lost opportunities due to inaccurate data.

Effective **risk mitigation** strategies prevent incidents that result in financial losses, legal liabilities, or reputational damage. By identifying and addressing risks before they materialize, CoEs can avoid costs associated with crisis management and system downtimes. Staying ahead of regulatory changes ensures compliance, avoiding penalties, fines, and costly legal proceedings. Proactively updating systems and processes to align with new regulations keeps the organization compliant without needing expensive, last-minute adjustments.

Avoiding future costs through **risk mitigation** requires current investments to maintain robust technology infrastructure to prevent future failures and security breaches. Proactive technology management, including regular updates and advanced security measures, avoids costs associated with data breaches, system outages, and obsolete technology. Effective vendor relationships and contract management avoid expenses related to poor service delivery, noncompliance with contract terms, or escalated pricing due to unfavorable contract conditions. Proactively managing these relationships ensures that contracts are transparent and beneficial, avoiding unexpected costs and disruptions.

Continuously **developing talent**'s skills and competencies in-house ensures the workforce can handle emerging challenges and technologies effectively, reducing the likelihood of costly errors and inefficiencies. This function helps avoid expenses related to underperformance, high turnover rates, and the need for external consultants or temporary replacements.

Cost savings

Cost savings have a multiplicative effect on net operating income because, assuming revenue remains constant, every dollar saved in operating expenses directly contributes to an equivalent increase in net operating income (NOI). This relationship is fundamental in business finance, where operating income is determined by subtracting operating expenses from revenue. Therefore, when costs are reduced through savings, these are not merely subtracted from expenses but are effectively transferred to the bottom line as increased operating income.

For instance, if a company generates $1 million in revenue and has $600,000 in operating expenses, its NOI would be $400,000. If the company implements cost-saving measures that reduce operating expenses by $100,000, the new operating expenses would be $500,000, and the NOI would increase to $500,000. Here, the $100,000 in cost savings directly amplifies the NOI by the same amount.

This multiplicative effect is particularly significant because increasing top-line revenue often requires proportional increases in costs or investments. In contrast, cost savings can be realized without necessarily increasing sales or output, making it a highly efficient way to improve profitability. Additionally, CoE management can often more directly control and implement cost savings than revenue-generating strategies, which more often are influenced by external market conditions and factors beyond the company's control.

Cost reduction

In the context of customer relationship management CoEs, several functions can easily be used to target cost reduction. This involves lowering costs but with the potential risk of reducing quality or value.

Budget function: This function focuses on controlling and cutting costs to stay within financial limits. Effective budget management involves scrutinizing current expenditures and identifying areas where costs can be reduced without significantly impacting the output or service quality.

Simplification function: Reducing complexity within CRM processes or systems can significantly reduce costs. Simplification can decrease the need for extensive training, support, and maintenance, directly cutting costs associated with these activities.

Service center function: Optimizing operations in service centers can significantly reduce the cost per interaction. This often involves improving process efficiencies, resulting in labor reductions or other operational expense cuts, achieving lower costs while maintaining service quality.

Change control function: By tightening change control processes, organizations can avoid costly errors and the need for redoing work. This function helps prevent unnecessary expenditures from correcting preventable mistakes, thus reducing overall costs.

Risk mitigation function: Proactively managing potential risks is crucial in avoiding expensive incidents, fines, or operational downtime. Effective risk mitigation reduces the likelihood of these costly events, directly decreasing potential expenses.

Cost-reduction per function

Here are detailed examples of how each function can reduce costs and the metrics to measure success.

The **goal-setting and prioritizing** function ensures precise resource alignment to high-impact projects, reducing wasteful spending on low-impact initiatives. This can be measured by tracking the percentage of projects meeting ROI targets.

Improved **transparency** reduces duplication and accelerates delivery times, which can measure the reduction in duplicated efforts across departments.

Focusing on the value creation function ensures **purpose-driven** spending, meaning it is directed toward initiatives that generate maximum returns. The increased cost savings for every value-driven initiative implemented can monitor this. Focused spending on **marketing** through precise targeting and customer segmentation can lower customer acquisition costs; metrics such as the decrease in cost per acquisition (CPA) reveal what's behind the success of this approach.

Streamlining **sales operations** reduces overhead and increases deal close rates, measured by the reduction in the cost of sales as a percentage of revenue. Efficient **service centers** can maintain customer satisfaction while lowering operational costs, tracked by decreased cost per service ticket.

Encouraging **ideation** leads to cost-saving products and processes, with the number of cost-saving ideas implemented as a metric. **Experimenting** with new methods and technologies can result in more cost-efficient operations, measured by the success rate of experiments that result in cost savings.

Investing in new products and services through the **value creation** function enhances revenue and reduces long-term costs, with ROI for new value creation initiatives providing a measure. Effective **talent management** optimizes workforce costs, measured by improving employee productivity rates.

Tight budget controls prevent cost overruns, with the percentage of projects completed within budget as a critical metric. A focused **product road map** avoids unnecessary feature costs, tracked by the number of features delivered on time without scope creep. Effective **change management** prevents costly disruptions, measured by reducing the expenses associated with unauthorized changes.

High **adoption** rates ensure full utilization of CRM investments, increasing system utilization rates and reducing support costs due to increased user self-sufficiency. The practices of an **Agile framework** minimize time to market and associated costs, with metrics such as the decrease in cycle time from development to deployment providing insight.

Streamlined workflows minimize inefficiencies, measured by time saved per process cycle. **Simplifying processes** cut unnecessary steps that consume resources, measured by reducing time spent training employees on processes. Improving system **maintainability** lowers long-term support costs, calculated by decreased maintenance costs over time.

Proactive **risk management** prevents costly incidents, borne out by metrics such as the reduction in financial impact from risk events. Robust **access controls** avoid costly security breaches, measured by the number of security incidents. Equally robust **continuity planning** reduces downtime costs, which are tracked by the decrease in downtime occurrences and recovery costs.

A **continuous development approach** prevents costly interruptions in employee activity and engagement, as measured by the reduction in labor costs. **Automated testing** processes decrease labor costs, as measured by the decrease in manual testing hours. **Developing talent** in-house reduces external training and recruitment costs, which metrics regarding external training expenses reveal.

Efficient **system integrations** reduce manual data-handling costs, which are tracked by reducing the number of person hours spent on data entry. **High-quality data** reduces costly decision-making errors, which are measured by reduced costs from data-related errors. **Standardized data** prevents costly data reconciliation issues, which are tracked by reducing the time spent on data cleanup.

By focusing on these cost-reduction strategies and their corresponding metrics, organizations can track the efficiency and effectiveness of investments in enhancing the functions of the CRM CoE. These measures validate cost-saving initiatives, ensuring they contribute to reduced costs and the organization's strategic value and operational excellence.

Cost optimization

Cost optimization reduces expenses and enhances operational efficiency, aligning spending with strategic goals to maximize value and support sustainable growth.

For example, suppose a company undertakes cost optimization initiatives utilizing AI to speed up coding. In that case, it might invest in a more efficient technology that reduces long-term operational costs, streamlines processes to eliminate redundancies, or renegotiates contracts to achieve better rates—all without compromising product quality or customer satisfaction. These actions decrease expenses and improve productivity and operational efficiency, leading to an increase in NOI.

The significant distinction here is that cost optimization can lead to increasing net operating income through the direct reduction of expenses and increasing revenue through improved operational efficiencies and customer experiences. This dual impact magnifies the effect on NOI because the organization saves money and potentially generates additional income through more effective operations. Thus, cost optimization contributes to a healthier bottom line by ensuring that resources are utilized most effectively, aligning spending with strategic priorities, and fostering an environment of continuous improvement and financial discipline.

For example, employing a financial metric to assess the CoE's efficiency, specifically the ratio of total CRM CoE expenditures to a metric for end-user adoption, provides a strategic marker for cost optimization efforts within the organization. If an organization allocates $100,000 annually towards its CoE, encompassing personnel, training, system enhancements, and governance, and the investment boosts end user adoption by 50% to 80%, this clearly demonstrates the fiscal benefit of strategic allocation of resources toward improved system utilization.

The effectiveness of this investment can be evaluated by calculating the cost per percentage increase in system adoption. As the investment remains fixed while adoption rates improve, the resulting ratio reveals a more cost-efficient use of resources per unit of increased adoption. This is particularly significant when the heightened user adoption correlates with measurable business improvements, such as enhanced sales productivity, faster customer service responses, or improved customer satisfaction ratings.

Measuring the CoE's performance showcases the CoE's commitment to cost optimization. It demonstrates how judiciously directed investments in CRM CoE enhance system adoption and operational efficiencies and contribute directly to the organization's financial health and competitive edge. Such a cost-effectiveness metric underscores the strategic value of CoE initiatives, providing a compelling case for ongoing or augmented investment.

Cost optimization strategies

In CRM CoEs, certain functions are particularly well suited to cost optimization, a strategy focused on maximizing resource effectiveness and efficiency without compromising quality or value. Here are examples.

Agile framework function. This methodology is designed to optimize resources and processes. By employing Agile practices, organizations can deliver value more quickly and efficiently, enhancing the overall effectiveness of the CRM system without compromising the quality of outcomes. For example, Agile methodologies enable teams to adapt to changes, prioritize high-value features, and continuously improve through iterative feedback, leading to faster delivery and better alignment with business needs. By fully embracing Agile practices, organizations can shorten development cycles, reduce time to market, and ensure that CRM systems are more responsive to evolving requirements, thereby justifying the cost through improved efficiency and quicker realization of benefits.

Workflow/process mapping function. Effective workflow or process mapping can significantly improve efficiency, resulting in direct and indirect cost savings. Streamlined processes reduce the time and resources required for execution without compromising service levels. For example, by identifying and eliminating redundancies, automating repetitive tasks, and optimizing resource allocation, organizations can achieve faster turnaround times and lower operational costs. Large companies might have business architecture teams within their departments. Still, it is worth questioning how many delivery teams fully comprehend and adhere to business architecture principles and processes while developing and implementing solutions. Ensuring that technical architecture teams collaborate with the business architecture team and align with the business architecture principals enhances cost efficiency by maximizing the return on the investment in process-mapping initiatives.

Talent development function. Investing in developing internal talent also optimizes costs. By enhancing employee skills and creating opportunities for career growth, there is less reliance on external consultants, which can be costly. Improved skills and knowledge within the team also boost retention rates, reducing the long-term costs associated with high turnover, such as recruiting and training new staff.

One significant challenge we encountered was the matrixed structure of CRM delivery teams. Technology product managers, for instance, may not have direct authority over their delivery team members. Developers, testers, and scrum masters might constitute one team, while administrators, architects, and business product managers, including business analysts, might be part of a different team. This structure hinders responding quickly to market demands. In this scenario, a CRM CoE can play a crucial role in devising a cross-team talent development strategy and coordinating its implementation.

Data quality and **data standardization functions.** Maintaining high-quality, standardized data across CRM systems enhances decision-making capabilities, operational efficiency, and effective customer targeting. This increases potential revenue through more precise targeting and improved customer relationships and prevents losses that might result from data errors. Thus, investing in data quality and standardization optimizes costs by reducing potential risks and enhancing revenue opportunities.

Testing and automation function. Automating repetitive tasks and robust testing protocols are crucial for optimizing workforce allocation. By automating routine and repetitive tasks, the labor costs associated with manual execution are reduced. Moreover, automation increases the accuracy and speed of processes, further optimizing operational costs and enhancing overall system reliability. Automation allows humans to focus on what they do best: solve complex problems.

Generative AI is changing CRM system testing processes and applications in several ways. Testing CRM applications substantially differs from traditional application testing like Java, .Net, and other cloud applications. CRM vendors like Salesforce frequently release new features and capabilities, leading to constant technical and functional changes. Without specialized testing platforms or experienced QA specialists, testing teams are challenged and become the bottleneck if the testing platform is not tuned to this fast change in platform architecture. A Center of Excellence can be crucial in identifying test solutions built to work well with cloud CRM platforms and make the testing process both faster and more reliable.

Generative AI, a disruptive technology, is revolutionizing the entire CRM delivery life cycle, but especially testing applications and processes. Generative AI can automatically create comprehensive test cases based on the application's requirements and user stories, significantly reducing the manual effort required from QA teams. It enhances test coverage by analyzing the entire application, identifying potential edge cases and areas that might have been overlooked, and ensuring broader and more thorough test coverage. AI systems can easily adapt and update test cases in response to frequent architecture changes and updates in CRM platforms like Salesforce, ensuring that tests remain relevant and effective without extensive manual intervention. Generative AI can optimize the execution of test cases by determining the most efficient order, identifying dependencies, and prioritizing critical tests to expedite the testing process.

Moreover, AI can analyze historical test data and application usage patterns to predict where defects are most likely to occur, enabling proactive testing and reducing the occurrence of critical issues. Natural Language Processing (NLP) capabilities allow AI to interpret and convert natural language into automated test scripts, streamlining the transition from requirements gathering to test execution. Generative AI can identify and correct broken test scripts caused by changes in the platform; for QA teams, this means reducing downtime and maintenance efforts. Additionally, AI can seamlessly integrate with continuous integration and continuous delivery and deployment (CI/CD) pipelines, enabling continuous testing and ensuring that CRM applications are consistently tested throughout the development lifecycle. By incorporating generative AI into testing processes, organizations can enhance efficiency, accuracy, and adaptability in their CRM application testing, ultimately leading to higher quality and more reliable software releases.

Examples of cost optimization per function

When enhanced through the lens of cost optimization, each function contributes to leaner operations, better resource utilization, and a more straightforward path to increased profitability. Below, we look at examples per function for optimizing costs with specific metrics to measure success.

Enhancing the **goal setting and prioritizing** function ensures that resources are allocated to high-impact projects, increasing ROI and saving costs from less-effective initiatives. Tracking ROI and the percentage of projects completed under budget that meet strategic goals can help measure this.

Improving **transparency** in decision-making and minimizing task redundancy leads to cost savings through better efficiency. This can be measured by reducing duplicate efforts and the need to redo work.

Ensure that every dollar spent is **purpose driven** so that it contributes to the value delivered to customers, which optimizes costs by eliminating wasteful spending. Metrics like increased customer lifetime value and decreased cost per lead help monitor this.

In the **marketing** function, using precise targeting and customer segmentation provided by the CRM can yield higher marketing ROI and reduce acquisition costs. Key metrics are evaluating cost per acquisition (CPA) and marketing qualified lead (MQL) conversion rates.

Streamlining **sales operations** can decrease the sales cycle length, reduce costs associated with customer acquisition, and increase sales conversion rates. Assessing the decrease in sales cycle time and the resulting increase in sales productivity reveals the efficacy of these improvements.

Optimizing **service centers** can reduce average handling times and increase first-contact resolution rates, lowering operational expenses. This can be tracked by measuring the reduction in cost per ticket and improvement in customer satisfaction scores.

Funding **ideation** leads to innovative cost-saving solutions and more efficient processes. The number of cost-saving initiatives generated versus those implemented serves as a powerful metric.

A budget for **experimentation** allows the organization to test and refine processes, leading to more cost-effective business practices. Monitoring the percentage of experiments leading to cost-saving measures and the time to implement changes helps measure this.

Enhancing the CoE's ability to create value can help the organization differentiate its offerings and potentially command premium pricing. Evaluating the increase in new revenue streams against investment in **value creation** is a valuable metric.

Effective **talent management** ensures that the right people are in the right roles, increasing productivity and reducing turnover-related costs. Assessing employee turnover rate and the cost savings from reduced hiring processes can measure this.

Tight **budget** control within the CoE can prevent cost overruns and ensure that spending aligns with strategic priorities. Monitoring budget variance and the percentage of initiatives within budget helps track this.

A clear **product road map** prevents scope creep and ensures development costs are invested in features that deliver the most value. Key metrics include tracking adherence to the road map and cost savings from prevented scope creep.

Robust **change control** reduces the risk of costly errors and ensures that changes yield the desired benefits. Evaluating the cost of change-related incidents before and after optimization helps measure this.

Ensuring high **adoption** rates maximizes the value of CRM investments and avoids the costs of under-utilized technology. Measuring system utilization rates and reducing support costs due to increased user self-sufficiency provides insight.

Practicing within an **Agile framework** can reduce development costs and time to market, leading to cost savings and quicker revenue generation. Assessing the cost of release cycles and cost reductions due to increased deployment frequency helps track this.

Optimizing workflows can eliminate inefficiencies, directly reducing operational costs. Monitoring the reduction in process cycle times and operational bottlenecks provides metrics for this.

Simplifying processes reduces complexity, which lowers training and maintenance costs. Evaluating the time saved or optimized per employee and the reduction in training costs measures this.

Improving **maintainability** reduces the long-term costs associated with system updates and downtime. Tracking the decrease in system downtime and maintenance costs helps measure this.

Minimizing risks through robust mitigation strategies can prevent potential future costs associated with data breaches or system failures. Monitoring the reduction in the financial impact of risk occurrences year-over-year helps track this.

Proper **access control** prevents unauthorized usage that could lead to costly security incidents. Measuring the decrease in security incidents and associated costs provides metrics for this.

Investment in **business continuity** ensures that the CRM system can quickly recover from disruptions, minimizing potential revenue loss. Measuring the reduction in recovery time after an incident and the associated cost savings provides insight into this.

A **continuous development** approach reduces the need for costly upgrades and maintains steady improvement at lower costs. Assessing the decrease in time between fixes and the reduction in emergency patching costs helps track this.

Automated testing reduces manual testing costs and human error, which improves efficiency. Tracking the reduction in manual testing hours and the associated labor costs and defects that end users find helps measure this.

Developing talent in-house can reduce the need for costly external consultants and build internal expertise. Measuring the reduction in external training costs and the increase in internal promotions helps track this.

Effective **integration** of the CRM system with other systems can reduce data silos and duplication of effort, saving on operational costs. Assessing the reduction in manual data entry hours and improved data processing time provides metrics for this.

High-quality data reduces errors and improves decision-making, potentially lowering costs from poor-quality information. Measuring the decrease in costs associated with data errors and improving decision-making speed helps track this.

Standardized data facilitates reporting and analysis, streamlines operations, and supports better business intelligence, ultimately leading to cost savings. Tracking the reduction in reporting errors and the time saved on data reconciliation provides metrics for this.

Optimization and reduction in tandem

Certain functions within the CRM CoE, such as purpose-driven, ideation, experimentation, and value creation, address cost optimization and reduction. These functions ensure that expenses directly contribute to creating value and optimizing expenditures efficiently. At the same time, they explore opportunities to reduce waste and unnecessary costs within the system. This dual approach helps save money and makes spending more effective and aligned with business outcomes.

The choice between cost optimization and reduction often depends on strategic intent. Functions that involve direct financial management, process efficiency, and risk management typically offer a more straightforward path to cost reduction. In contrast, functions focusing on methodologies, talent development, and data quality provide broader opportunities for cost optimization, enhancing overall operational efficiency and effectiveness. This strategic distinction can help organizations decide the best approach to cost management based on their specific operational needs and financial goals.

Top-line revenue justifications

Shifting from the conventional cost-center perspective, CRM Centers of Excellence are being reimagined as strategic assets that drive revenue and enhance customer satisfaction. By adopting innovative methodologies, CRM CoEs can transform the focus of functions to long-term sustainable business growth, thus optimizing costs, improving Net Promoter Scores, and fueling top-line growth.

In recognizing the potential of CRM CoE functions to impact revenue and NPS positively, we advocate for enhancements rooted in customer-centric strategies. By strategically investing in functions such as marketing, sales operations, and service centers, we directly influence customer perceptions and engagement, which are pivotal drivers of loyalty and revenue. For instance, honing the precision of marketing campaigns with AI can boost conversion rates while streamlining sales operations shortens the sales cycle, contributing to a robust increase in revenue. Similarly, fortifying service center efficiency with AI-enabled processes enhances customer satisfaction, a cornerstone of customer retention and a predictor of sustained revenue.

Metrics such as conversion rates, sales revenue growth, and first-contact resolution rates provide quantifiable evidence of progress and underscore the direct correlation between function enhancements and financial outcomes. These enhancements are not one-off improvements but are sustained through continuous learning and feedback loops, ensuring the CoE remains agile and responsive to evolving customer needs and market trends. This dynamic approach secures the CoE's position as a catalyst for revenue generation, dispelling the outdated notion of IT as merely a cost center.

The CRM CoE emerges as a proactive participant in shaping an organization's financial and competitive landscape. The focus shifts from governance to active engagement in driving business success. The outcome is a CoE that justifies its expenditures through cost savings and validates its investments through measurable contributions to customer loyalty and revenue growth.

This approach leverages CRM functions as strategic assets, directly influencing customer engagement and satisfaction, crucial for revenue growth and NPS improvement.

Below are justifications and metrics for enhancing specific CoE functions that are particularly effective in driving top-line revenue growth.

Enhancing targeting and personalization in **marketing** campaigns can increase conversion rates and the acquisition of customers. Streamlining the **sales** process reduces the sales cycle duration, allowing for more transactions and faster revenue generation. Metrics such as revenue growth and reduced average sales-cycle length illustrate these improvements.

Improving **service center** efficiency boosts customer satisfaction and retention, directly influencing repeat business and referrals. Metrics like first-contact resolution rates and customer retention rates provide evidence of these enhancements. **Experimentation** and **innovation** associated with developing new products and services that meet evolving customer needs drives market differentiation. It opens new revenue streams, with revenue generated from new products or services as a critical metric.

Efficient **integration** of CRM systems enhances customer data analysis, leading to better customer experiences and increased upsell opportunities. The increase in upsell and cross-sell revenue is a measurable outcome. High-quality, reliable **data** allows for more accurate targeting and customization of offers, enhancing customer response rates and satisfaction; in this case, increased revenue per customer is the metric.

Ensuring high **adoption** rates of CRM tools across the organization enhances employee efficiency and customer interaction quality, leading to better sales outcomes. Metrics like revenue-per-sales-representative demonstrate this impact. Encouraging a culture of **innovation** can lead to developing unique solutions that address specific customer needs, differentiating the brand and boosting sales. A crucial metric is the number of new customer accounts resulting from innovative solutions.

Implementing **Agile frameworks** in project management and product development ensures that services and products are rapidly adjusted to market demands, keeping the organization competitive. The time to market for new features and products is a critical metric. Systematic collection and analysis of customer feedback inform product enhancements and service improvements, leading to increased customer satisfaction and loyalty. An increase in the **Net Promoter Score** is the measurable evidence.

Standardizing data formats avoids costly data transformation projects, with reduced expenditures on data conversion serving as the metric.

These revenue-focused justifications and corresponding metrics demonstrate how CRM CoE functions can drive business success. This transformation from cost centers to revenue drivers marks a significant evolution in the perception and impact of CoEs, positioning them as indispensable players in shaping the financial and competitive landscape of business costs.

Conclusion: Revenue-centric stewardship

Chapter 4 provides a framework for optimizing the economic aspects of CRM CoE functions. Organizations can achieve financial benefits by focusing on cost avoidance, savings, reduction, and optimization while enhancing CRM capabilities. Cost management strategies directly increase NOI, support sustainable growth, and prevent unnecessary expenditures. Aligning CRM CoE initiatives with broader organizational financial goals ensures that CRM CoEs drive growth, enhance customer experiences, and foster innovation.

This chapter introduces revenue generation as a critical piece of the financial discourse within a CRM Center of Excellence. We go beyond cost considerations, advocating for a revenue-centric approach that aligns financial strategies with the potential for sustainable growth and profitability.

Cost savings, while traditionally about prudent fund use, are reenvisioned here as investments that elevate the organization's financial standing without compromising service delivery. Cost reduction, often associated with cutting corners, is reframed as a resizing that streamlines operations without impacting core functionalities vital for sustained revenue streams. Most significant, cost optimization is elevated from a simple balancing act to a deliberate strategy that positions every dollar spent as an engine for value generation and revenue growth.

By intertwining these financial strategies with the CRM CoE's operational agenda, leaders are empowered to champion functional enhancements that deliver an immediate fiscal impact and pave the way for future income. The strategic alignment of these economic principles with CoE functions ensures that each enhancement is an investment in the organization's long-term profitability, competitive edge, and viability. As such, this chapter lays the groundwork for CoEs to justify their expenditures and secure funding by demonstrating their value to an organization with measurable contributions to cost efficiency and revenue growth.

CHAPTER 5
The Cost of Running the CRM Ecosystem

Executive summary

This chapter provides an example of a cost analysis for operating a CRM ecosystem, focusing on Salesforce Health Cloud for an organization with 1,000 users. Understanding these financial implications is critical for companies considering or currently using this robust platform. This analysis—again, an example—gives you the necessary information to analyze and reflect on CRM investment decisions in a specific context.

In this example, we calculate the annual cost of running a Salesforce Health Cloud ecosystem for 1,000 users utilizing industry-standard estimates. We analyze the revenue required to support these costs, providing a clear picture of the financial health needed to sustain such an investment. In addition, we explore methods for calculating the Return on Investment (ROI) and the payback period, including industry benchmarks and success stories to illustrate effective CRM investment management. Last, we discuss potential risks and mitigation strategies to ensure a reliable and efficient CRM operation. This chapter aims to equip readers with a thorough understanding of the cost components of running a CRM ecosystem like Salesforce Health Cloud, aiding informed decision-making regarding investing, expanding, or optimizing this crucial business tool.

This chapter examines the hypothetical cost analysis of operating a CRM ecosystem, explicitly focusing on Salesforce Health Cloud for an organization with 1,000 users. Understanding the financial implications and ensuring value realization from CRM investments[33] is critical for companies considering or currently using this robust platform. We provide a detailed breakdown of all costs of running such a system. These include:

- **Implementation costs.** Initial expenses for setting up and deploying the CRM system within the organization's IT infrastructure.

- **Licensing fees.** These are the direct costs to Salesforce for user access to the Health Cloud environment.

- **Customization and integration.** The costs associated with tailoring the system to the organization's specific needs and integrating it with other systems.

- **Maintenance and support.** Ongoing expenses for maintaining the system's functionality and receiving support from Salesforce or third-party providers.

- **Training and adoption.** Investments in training staff to effectively use the CRM and strategies to encourage its adoption across the organization.

- **Scaling and upgrades.** Costs involved in scaling the solution to accommodate growth and upgrading the system to incorporate new features and capabilities.

- **Additional data and storage.** Expenses for extra storage, including data backups and security measures, exceed the base licensing fees.

- **Miscellaneous costs.** Other costs that might not neatly fit into the other categories but are essential for the comprehensive operation of the CRM ecosystem.

Industry-standard guesstimates

Using industry-standard estimates, we will calculate the annual cost of running a Salesforce Health Cloud ecosystem for 1,000 users. We will then analyze the revenue required to support these costs, providing a clear picture of the financial health needed to sustain such an investment.

In addition, we will explore methods for calculating the Return on Investment (ROI) and the payback period. This includes comparing industry benchmarks and sharing success stories to illustrate how organizations have effectively managed and justified their CRM investments. Last, we will discuss potential risks associated with such a system and strategies for mitigating these risks to ensure a reliable and efficient CRM operation.

---

33   Glen S. Gooding, *The IT Financial Management Lifecycle: Budgeting, Costing, Chargeback, and Benchmarking* (London: Kogan Page, 2010).
Todd Tucker, *Technology Business Management: The Four Value Conversations CIOs Must Have With Their Businesses* (Bellevue, WA: TBM Council, 2016).
Sanjeev Purushotham, *The Economics of IT Cloud Computing: Billing, Capacity, and Costing* (New York: Springer, 2018).

By the end of this chapter, readers will have a good understanding of the cost components involved in running a CRM ecosystem like Salesforce Health Cloud. This knowledge will aid in making informed decisions about investing in, expanding, or optimizing this crucial business tool. As you compare these numbers to your own experience and situation, consider how the use of AI changes the costs of implementing or maintaining a system.

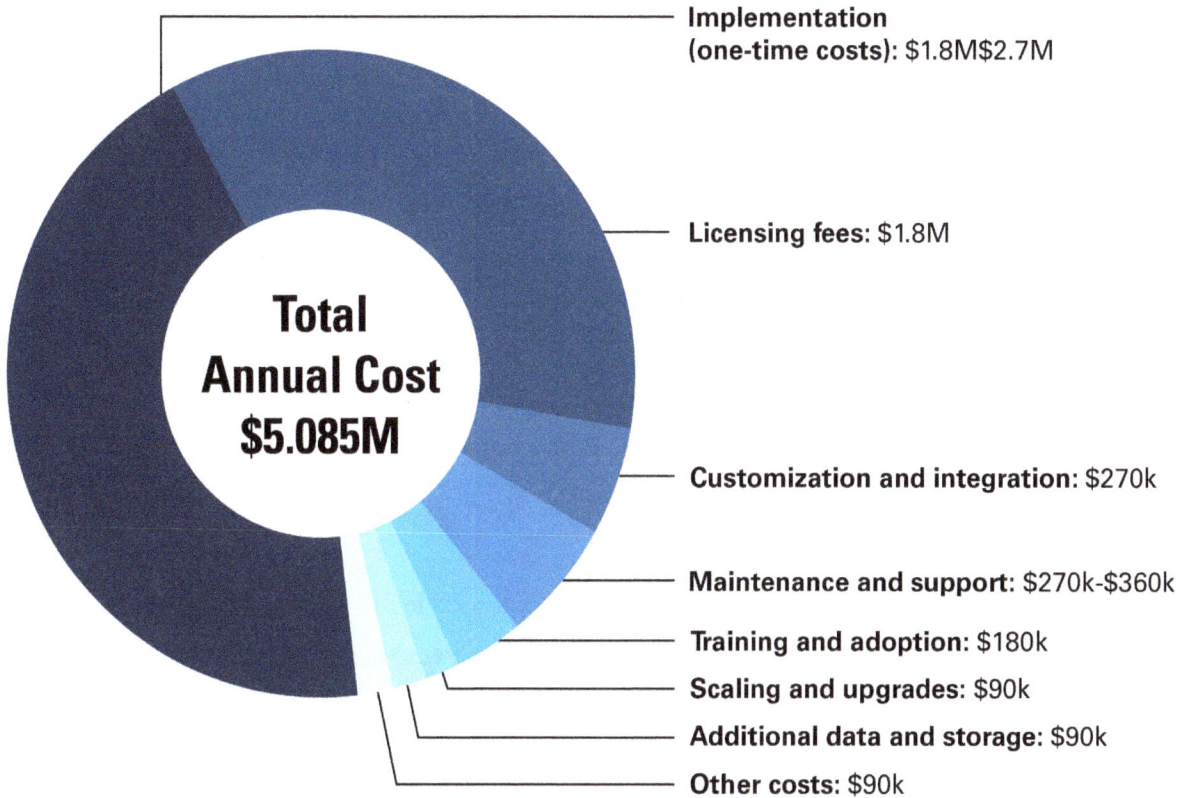

Implementation (one-time costs): $1.8M$2.7M

Licensing fees: $1.8M

Total Annual Cost $5.085M

Customization and integration: $270k

Maintenance and support: $270k-$360k

Training and adoption: $180k

Scaling and upgrades: $90k

Additional data and storage: $90k

Other costs: $90k

Figure 41: Cost breakdown (in US dollars) of running a CRM ecosystem.

Costs of running a CRM ecosystem

One-time implementation costs: Implementation involves initial setup fees and costs for consulting services. These costs typically range from 1 to 1.5 times the annual licensing fees, which can total from $1.8 million to $2.7 million.

Licensing fees: Licensing fees are a significant expense for an organization with 1,000 Health Cloud users. At $150 per user per month, the annual cost amounts to $1.8 million.

Customization and integration: Tailoring the CRM to meet specific organizational needs involves costs for custom development and integration with existing systems. This is estimated at 15% of the annual licensing fees and would cost $270,000.

Maintenance and support: Ongoing maintenance and support costs, including salaries for Salesforce administrators and additional support plans, range from 15% to 20% of the annual licensing fees, or $270,000 to $360,000.

Training and adoption: The costs associated with training programs and change management activities are crucial for ensuring effective system adoption. They are estimated at 10% of the annual licensing fees, or $180,000.

Scaling and upgrades: As the organization grows, additional costs are incurred for more licenses and system upgrades. This is 5% of the annual licensing fees, or $90,000.

Additional data and storage: The cost of extra data storage is estimated at 5% of the annual licensing fees, or around $90,000, depending on actual data usage.

Other costs: Miscellaneous costs, including mobile access and compliance with regulations, also account for about 5% of the annual licensing fees, totaling $90,000.

Total annual cost: The total annual cost for operating a Salesforce Health Cloud ecosystem for 1,000 users is approximately $5.085 million. This includes all the costs, from licensing to miscellaneous expenses.

Revenue needed to support the cost: To cover the annual cost of $5.085 million with a profit margin of 10%, the required revenue would be approximately $50.85 million. This calculation is based on general industry conditions and should be adjusted according to specific company needs and profit structures.

Return on investment (ROI) and justification: To determine the viability of an investment in Salesforce Health Cloud, it is imperative to conduct a comprehensive analysis of both quantifiable and qualitative benefits. These benefits may include increased sales, enhanced customer retention, efficiency gains, reduced labor expenses, and improved customer satisfaction. It is necessary to quantify these benefits in monetary terms in order to calculate the ROI using the following formula:

$$\textbf{ROI = ((Total Benefits Total Costs) Total Costs)) x 100\%}$$

It is crucial to consider the payback period, which refers to the duration required for the benefits to offset the costs associated with the CRM investment. Large enterprises often utilize complex algorithms to calculate the payback period. Furthermore, industry benchmarks, case studies, and success stories can be leveraged to strengthen the justification by showcasing similar successful implementations.

Additional variables for factoring costs

1. Scaling and upgrades: As the CoE matures, its needs regarding scalability and system upgrades may evolve. It might focus on establishing a robust foundational system early in its lifecycle. Still, as the organization grows and the CoE becomes more integrated into the company's core processes, the need for more advanced features and integrations can increase, potentially raising costs.

2. Customization and integration: Initially, customization and integration costs might be high as the CoE works to tailor the CRM system to the organization's specific needs. Over time, these costs might decrease as these systems stabilize and the CoE gains more experience. However, they could also rise again if new business needs dictate significant changes or ongoing innovation is prioritized.

3. Maintenance and support: In the early stages, maintenance and support costs may be higher due to the learning curve associated with new technology and the need to address initial system bugs or shortcomings. As the CoE matures, these costs stabilize as staff become more proficient and systems are optimized. However, ongoing innovation or expansions can again increase these costs.

4. Training and adoption: Early on, significant investment in training and adoption is critical to effectively using the CRM system. As the CoE matures, the foundational training costs may decrease, but ongoing training related to new features or for new employees might be needed, which could vary the costs.

5. Data and storage costs: As more data accumulates and the company relies increasingly on data-driven insights, additional storage and data management costs will increase. However, more mature CoEs might also find more efficient ways to handle data or negotiate better rates with providers.

6. Other costs: Miscellaneous costs such as regulatory compliance or mobile access needs might evolve with the CoE's maturity. A mature CoE might have streamlined processes that reduce some costs but could incur new costs as it expands its scope or faces new regulatory requirements.

7. Revenue requirements and ROI: As the CoE matures, it might drive higher efficiency and effectiveness, improving the ROI of the CRM investment. A mature CoE can better demonstrate the value of the CRM system through enhanced customer relationships and more strategic data usage, which can justify the costs and support higher revenue requirements.

Overall, the maturity of the CoE often leads to better forecasting, budget management, and cost optimization as the organization becomes more experienced in managing its CRM ecosystem. However, this evolution can also bring new challenges and costs as expectations and objectives evolve.

Industry references and benchmarks

There are not many industry references for estimating cost of ownership for a CRM CoE. They tend to focus on CRMs, not the CoE. And they tend to be generic, not specific to a particular industry. The following may be helpful in getting a comparative sense of costs and return.

- A Salesforce ROI study[34] by Nucleus Research found an average ROI of 348% with an 8.8-month payback period.

- Gartner's total cost of ownership analysis[35] shows a range of $70 to $210 per user per month, depending on features.

- Forrester's Total Economic Impact Study[36] reported a 292% ROI and less than six months payback for 1,000 users.

- IDC analysis of customer relationship management ROI[37]: Average ROI of 245% over five years with a 7.5-month payback period.

These studies and benchmarks provide the backdrop to support the cost estimates and ROI analysis, illustrating that while substantial, the costs align with the potential returns in similar settings. And, yes, the costs and considerations outlined above can significantly change as the Center of Excellence evolves.

Conclusion

Chapter 5 provides a detailed cost analysis of operating a Salesforce Health Cloud ecosystem for 1,000 users, outlining all cost components. Understanding these costs is crucial for aligning CRM initiatives with strategic goals to drive growth, enhance customer experiences, and foster innovation. By strategically managing costs and continuously aligning CRM initiatives with business objectives, organizations can make informed decisions about their CRM investments, enhance customer relationships, drive innovation, and achieve long-term success.

---

34   Nucleus Research, "Salesforce ROI Case Study: Pearson Education," June 15, 2021 - ROI Case Studies V91, https://nucleusresearch.com/research/single/salesforce-roi-case-study-pearson-education/

35   Gartner. "Use Total Cost of Ownership to Optimize Costs and Increase Savings." Last modified January 19, 2018. https://www.gartner.com/en/documents/3847267.

36   Forrester Consulting, The Total Economic Impact™ of GE Digital's Asset Performance Management, commissioned by GE Digital, July 2022, https://www.ge.com/digital/sites/default/files/download_assets/forrester-total-economic-impact-ge-digital-apm.pdf

37   IIDC, "ROI from CRM Ranges from 16% to 1000%," The Wise Marketer, accessed November 14, 2024, https://thewisemarketer.com/roi-from-crm-ranges-from-16-to-1000/

CHAPTER 6
Center of Excellence Leadership

Executive summary

Chapter 6 underscores the pivotal role of the CRM Center of Excellence (CoE) leader in driving company growth, customer satisfaction, innovation, and value realization as a result of CRM initiatives. The CoE leader[38] ensures CRM initiatives align with business strategies, fostering collaboration and continuous improvement. Key responsibilities include strategic alignment, executive engagement, dismantling silos, change management, program oversight, talent development, and operational efficiency.

The leadership team evolves as a CRM CoE transition from concept to implementation and maturity, incorporating roles such as initiators, funders, sponsors, evaluators, and the CoE Lead. This chapter outlines how CoE leaders establish standards, ensure governance and compliance, manage change, promote innovation, facilitate cross-functional collaboration, provide user support, measure performance, and manage resources. Additionally, it emphasizes the importance of strategic relationships with CRM technology platform providers, optimizing CRM resources, and navigating multiple platforms to enhance customer engagement. The ideal CoE leader combines strategic leadership, technical expertise, interpersonal skills, and an innovative mindset.

---

38   Simon Sinek, *Leaders Eat Last: Why Some Teams Pull Together and Others Don't* (New York: Portfolio, 2014).
Patrick Lencioni, *The Five Dysfunctions of a Team: A Leadership Fable* (San Francisco: Jossey-Bass, 2002).
Brené Brown, *Dare to Lead: Brave Work. Tough Conversations. Whole Hearts.* (New York: Random House, 2018).

The leadership team evolves as the CRM CoE ecosystem transitions from concept to implementation to maturity. While the role of CRM CoE leader is more constant, the composition of the leadership team varies by stages. We see possible members of a CoE team this way:

Initiators
Individuals who identify the need for a CRM CoE and propose its creation to enhance customer relationship management within the organization.

Funders
Those who secure the financial resources to ensure the CRM CoE can function effectively.

Sponsors
Typically, senior executives who champion the CRM CoE within the company, securing organizational buy-in and providing the visibility and authority needed for change implementation.

Evaluators
People responsible for assessing the CRM CoE's performance and impact, monitoring outcomes, analyzing data, and suggesting improvements.

Leaders
They manage the strategic direction and daily operations, aligning the CRM CoE's initiatives with the company's overall goals and ensuring the team is well-supported and successful.

Figure 42: Leadership team roles evolve.

Strategic CoE leadership tasks

Strategic leadership tasks may be performed by different team members at different stages of the CoE's development, but as the CoE evolves, they are likely to truly become team tasks guided by the CoE leader.

Strategic alignment. CoE leaders ensure CRM initiatives perfectly align with the company's business strategies and priorities, seamlessly connecting its vision with its operational execution.

Vision and strategy. CoE leaders articulate and actively translate a broad vision into actionable strategies. Setting clear, strategic objectives ensures that every facet of the CRM initiative leverages technology targeted to drive business growth and improve customer relationships. This vision might involve integrating CRM systems with other digital tools to foster data accessibility and customer insights, cultivating a data-driven culture within the organization.

Executive engagement. They are pivotal in securing executive buy-in and the necessary resources, positioning CRM as a central element of the company's future success.

Initiators, funders, and sponsors. These roles are crucial in the early stages. They identify the need for a CoE, secure financial resources, and champion the CoE within the organization to ensure it has the necessary backing and visibility for successful implementation.

Silo dismantling. These leaders face the challenge of breaking down organizational silos, advancing an environment of teamwork and a culture of innovation and unified purpose.

Cross-functional collaboration. The CoE leader plays a critical role in breaking down silos and encouraging cross-departmental collaboration to ensure CRM strategies are effectively implemented and aligned with overall business objectives. This involves coordinating with IT, marketing, sales, and customer service teams to ensure CRM systems serve broad-based business needs and drive unified strategies. Their efforts in promoting collaboration contribute significantly to a holistic view of the customer journey, elevating overall customer satisfaction.

Change catalyst. As drivers of adoption and effective change management, CoE leaders ensure CRM tools and strategies are embraced across the organization, transforming business operations. Evaluators play a crucial role here, continuously assessing performance and suggesting improvements to ensure ongoing alignment with strategic objectives.

Change management. Effective change management under the CoE involves preparing the organization for new CRM systems and updates, facilitating training sessions, and managing resistance to change. CoE leaders develop comprehensive change management plans that include stakeholder analysis, communication strategies, and feedback mechanisms to ensure smooth transitions and high adoption rates. Their proactive approach helps minimize disruptions to business operations and nurtures user competence and acceptance of new CRM functionalities.

Championing customer centricity. CoE leaders ensure that customer centricity is at the heart of all CRM initiatives, providing strategies and actions that consistently enhance customer experience and value.

Operational CoE leadership tasks

As with strategic leadership tasks, operational leadership involves a team of leaders with the CoE sitting in the hub. Especially within larger CoEs, the internal leadership will divide up responsibilities for monitoring and collaborating with leadership of related groups to drive these operational tasks.

Operational efficiency. They ensure CRM system delivery operations are efficient, cost-effective, innovative, and deliver value to stakeholders.

Program and project oversight. They manage CRM-related programs and projects, steering them towards timely and successful completion.

Innovation and continuous improvement. CoE leaders champion innovation by continuously seeking opportunities to enhance CRM systems and processes. This could involve piloting new CRM features, integrating advanced analytics for deeper customer insights, and exploring artificial intelligence capabilities to automate routine tasks and personalize customer interactions. By fostering a culture of innovation, they keep the organization at the forefront of CRM technology developments, ensuring it remains competitive in a rapidly changing business environment.

Talent development. Developing people and enhancing CRM competencies across teams is crucial, equipping the organization to fully leverage CRM capabilities. AI poses unique challenges to the development of CRM systems, including disrupting the traditional training path from a business analyst or developer to more strategy-oriented professional or managerial roles. How can staff acquire the lower-level experience needed to form wise, experience-based decisions in strategic roles like architect or product manager when AI handles the lower-level tasks? As AI is incorporated into everyday operations, companies must determine how to maintain the experience component of talent development.

Resource management. Within the CoE, resource management involves strategically allocating human and financial resources to support CRM initiatives effectively. Leaders are tasked with balancing the budget, assigning staff to CRM projects based on their expertise, and ensuring that CRM investments align with the organization's strategic priorities.

Foundational CoE leadership tasks

The tasks of foundational leadership fall more directly under the control and responsibility of the CoE. Which ones remain under the control of other groups in the company will vary by company organizational structure. All foundational tasks are of vital and immediate interest to the CoE leadership.

Foundation management. Leaders create and maintain standards, procedures, and best practices to keep the company's CRM technology platform secure, resilient, and compliant.

Governance and compliance. The CoE leader implements robust governance frameworks to oversee the operation and use of CRM systems across various departments, ensuring that all CRM activities adhere to internal policies and external regulations. This involves regular audits, compliance checks, and updates to the governance framework in response to new regulatory demands. Effective governance helps mitigate risks associated with data breaches and noncompliance penalties, which is important to protecting the organization's reputation.

Standards and procedures. Under the leadership of the CoE, standardized procedures and best practices are developed to ensure uniformity across all CRM-related activities. This includes standardizing data entry processes to reduce errors, aligning customer interaction strategies across channels to provide a seamless customer experience, and establishing clear guidelines for CRM usage that comply with regulatory requirements. These standards are crucial for maintaining the integrity and security of data, which are required for customer trust.

User support and training. To maximize the impact of CRM initiatives, CoE leaders ensure ongoing user support and up-to-date training. They oversee the creation of comprehensive training programs tailored to different user roles and learning styles, set up dedicated support channels, and continually maintain a feedback loop to refine training materials and support services. Investing in user education increases operational efficiency and empowers employees with the skills needed to leverage CRM tools effectively.

Scalability and sustainability. CoE leaders ensure CRM systems are both scalable and sustainable, capable of adapting to future growth and technological advances. This requires forward-looking planning, regular system upgrades, and integration of scalable cloud-based solutions that can accommodate increasing data volumes and expanding business activities.

Performance measurement. CoE leaders establish a system of performance metrics to gauge the success of CRM initiatives, reflecting vital business goals. These metrics include customer retention rates, lead conversion rates, and customer satisfaction scores. Regular monitoring and reporting on these metrics allow leaders to make data-driven decisions to optimize CRM strategies and operations.

Leveraging the platform ecosystem. Consider the Salesforce ecosystem, known for its robustness and comprehensive suite of integrations and applications available through AppExchange. This CRM platform has a broad range of functionalities that cover various aspects of customer relationship management, from sales and customer service to marketing automation and business analytics. For many organizations, a platform like Salesforce can be a single, full-bodied platform that meets all their CRM needs.

However, a strategic CoE leader must assess whether transitioning to a unified platform like Salesforce could benefit the organization. This decision involves understanding the depth and breadth of the platform's capabilities and determining if it can effectively replace other systems. By consolidating onto a platform, the organization can reduce the complexity of its CRM landscape, streamline operations, and enhance the overall user experience with a more integrated approach.

Navigating multiple platforms. Organizations often find themselves managing multiple CRM platforms to meet different business needs. This can lead to inefficiencies, increased costs, and complexity in data management. For example, an organization might use one CRM platform for its sales team and another for customer service operations. This separation can create data silos, complicate the user experience, and make it difficult to achieve a unified view of customer interactions.

The CoE leader's role involves evaluating whether multiple platforms genuinely serve the organization's best interests or if consolidating onto a more comprehensive platform could yield better results. This evaluation must consider the costs of licensing, training, and maintaining systems versus potential gains in efficiency, productivity, and customer satisfaction.

Strategic partnership with platform provider. One of the most strategic roles of a CRM CoE leader is managing the relationship with CRM technology platform providers. This relationship is critical, particularly in environments where multiple CRM platforms are in use since it directly impacts the organization's ability to streamline operations and maximize the efficiency of CRM initiatives. Using the example of the Salesforce platform, CoE leaders should:

- **Participate in development and customization.** Work with Salesforce to tailor the platform to the organization's needs. This could involve custom application development or configuring the platform's features to better suit business processes.

- **Engage in proactive planning.** Regularly review and plan for new platform features and updates. Understand how these can be harnessed to improve business operations or enter new markets.

- **Training and certification.** Encourage team members to participate in platform training and certification programs. This ensures the organization fully utilizes all platform features and stays updated on best practices.

- **Feedback and collaboration.** Provide input to the platform vendor on product functionality, which can help shape future developments. Participating in beta tests of new features or joining platform customer advisory boards can place the organization at the forefront of CRM technology.

- **License and product count management.** Actively manage the number of licenses and products used within an ecosystem to optimize cost and efficiency. Regularly assess needs against actual usage and adjust accordingly to avoid unnecessary expenditure.

- **Optimize production instances and sandbox environments.** Evaluate the number of CRM production, development, and testing instances to streamline operations and reduce complexity. Consolidate where possible to minimize overlap and redundancy, ensuring each instance serves a distinct and necessary purpose.

- **Address platform overlap and redundancy.** In cases where multiple platforms (Salesforce.com, Microsoft Dynamics, SAP, Oracle CRM on Demand, etc.) are still necessary, the CoE leader plays a crucial role in managing overlap and redundancy. They must establish clear boundaries and guidelines for when and how each platform should be used to avoid duplication of effort and to ensure data consistency across systems. Regular audits and reviews can help identify overlapping functionalities and provide a basis for rationalizing the CRM landscape.

By managing these aspects effectively, the CoE leader ensures that the organization's relationship with its CRM technology provider(s) supports strategic business objectives, leverages technological advancements, and drives continuous improvement in customer relationship management. This strategic partnership, particularly with the platform provider, can transform the organization's CRM landscape, making it more streamlined, integrated, and effective at meeting customer needs.

The CRM Center of Excellence leader is pivotal in orchestrating the dynamic interplay between strategic vision, operational execution, and technological innovation within an organization. By ensuring CRM initiatives align with business strategies, fostering executive engagement, and dismantling silos, the CoE leader lays the groundwork for a transformative impact on customer satisfaction and company growth.

Operational tasks ranging from managing change, nurturing talent, and ensuring compliance to championing customer centricity are all geared towards creating a resilient, efficient, and customer-focused CRM ecosystem. These efforts ensure that the CoE's strategies are deeply integrated into the organizational fabric, enhancing overall performance and customer engagement.

Moreover, strategic relationship management with platform providers showcases the CoE leader's critical role in leveraging technology to meet and exceed organizational goals. The leader ensures that CRM technology is a robust foundation for business operations through optimized use of CRM resources, streamlined operations, and a unified platform approach, where applicable.

This leadership ensures that CRM technology is not merely an operational tool but a strategic asset that propels the organization forward in a competitive and rapidly evolving business environment. The journey of the CRM CoE is one of continuous adaptation and improvement, aiming to position the organization at the forefront of excellence in customer relationship management. Each step moves toward better integration, more effective operations, and a stronger focus on the customer.

Profile of the ideal CoE leader

Selecting a leader for the Center of Excellence who embodies the company culture is crucial for the success of customer relationship management initiatives. A leader aligned with the organization's values facilitates smoother adoption of CRM strategies, fosters collaboration, and ensures that technological implementations resonate with the team's core behaviors and beliefs. This cultural fit enhances team morale, eases change management, and drives a unified commitment toward customer engagement objectives. As the CoE matures, the leader's profile must evolve to meet the organization's and the industry's dynamic demands.

Does the CoE leader need to come from within the CRM?

Not necessarily. While having a CoE leader with extensive CRM experience can be beneficial, it is not a strict requirement. The essential qualities of a successful CoE leader include strong leadership skills, a strategic vision, and the ability to drive cross-functional collaboration. One of the critical responsibilities of a CoE leader is aligning customer relationship management initiatives with the organization's core mission and strategic goals. This strategic alignment, coupled with their proficiency in managing change, fostering innovation, and ensuring effective governance, sets them apart. Hence, leaders with a background in business strategy, technology management, or project management can excel in this role, provided they understand the strategic importance of CRM and are committed to leveraging it for enhanced business outcomes.

To avoid conflicts between delivery and governance objectives, we recommend that the Center of Excellence team retain responsibility for a set of functions, not including actual product or project delivery. The CoE team should retain the following: CRM strategy, architecture, governance, administration, security, compliance, advisory services, and other common tasks, such as release management and CRM data governance. By assigning these responsibilities to the CoE team, companies can ensure that there is a clear separation of roles and that the CoE team can focus on its core mission of driving innovation and improving the overall quality of IT services.

In our experience, a common pitfall is assigning a delivery leader or an enthusiastic individual as the leader of the CRM Center of Excellence solely based on their eloquence and self-proclaimed expertise. This approach often overlooks crucial qualities and skill sets necessary for effective CoE leadership.

Leadership evolution

As CoE leaders gain experience and confidence, they typically undergo changes affecting style and substance. These should affect actions and decision-making in the following areas:

Change management. The early stages require setting up structures that support adopting new technologies. In later stages, the leader focuses on sustaining change, making iterative improvements, and embedding resilience to ongoing changes in technology and market dynamics.

Data-driven decision making. Starting with leveraging data for strategic decisions, this competency must evolve into advanced data analytics, predicting customer behaviors, and personalizing customer interactions to promote engagement and satisfaction.

Team development and collaboration. Initially, the focus is on building and managing effective teams. As the CoE grows, the leader should focus on fostering a culture of innovation, cross-departmental collaboration, and integrating CRM practices across the organization.

Educational background. A degree in a relevant field is fundamental. As the CoE evolves, continued education through advanced degrees or certifications in emerging CRM technologies and strategies will be necessary.

Professional experience. Starting with a solid background in CRM, the leader must continually expand their expertise to include innovations in CRM and customer experience, staying abreast of industry changes and technological advances, particularly in AI.

Innovative and growth mindset. An openness to learning and innovation is crucial. The leader must continue to champion in-house innovation, driving growth through new CRM technologies and strategies.

Problem solving. The leader tackles setting up structures and solving initial implementation challenges early on. As the CoE matures, problem-solving becomes more about navigating complex business landscapes and ensuring that CRM strategies align with global business objectives.

Solution architecture. The leader must initially ensure scalable and user-friendly CRM solutions. As technology advances, they must oversee the integration of new technologies, ensure that staff is fluent in their use, and ensure these solutions continue to meet evolving business needs.

Executive presence and influence. Strong communication and influence skills are essential to engaging stakeholders from the outset. As the organization evolves, these skills become crucial in advocating for strategic shifts and securing buy-in for new directions.

Emotional intelligence. High emotional intelligence is vital for navigating challenges. Over time, this skill must adapt to managing more complex stakeholder relationships and fostering a culture of empathy and adaptability.

Early establishment and talent management. Initial priorities include quickly establishing the CoE and managing talent. Later, to retain top talent and align expertise with business needs, it will be necessary to help map out career paths for in-house talent and adapt team structures to fit their evolving CRM roles.

Delivering and measuring outcomes. Initially focused on tangible outcomes to demonstrate CRM's value, the leadership must evolve to focus on long-term strategic benefits, such as customer lifetime value and integration of CRM into overall business strategy.

Community involvement and thought leadership. Starting with engagement in CRM communities and forums, the role evolves into a thought leadership position, influencing industry standards and practices and bringing innovative ideas into the organization.

Balancing delivery and governance. The early focus on achieving a balance of governance and delivery shifts towards a more strategic focus, ensuring that CRM practices align with broader organizational goals and adapting governance models to accommodate growth and complexity.

This evolutionary approach ensures that the Center of Excellence leader remains practical and relevant as the organization's CRM capabilities mature. By preparing for these evolving demands, the leader can drive sustained success, align CRM initiatives with strategic business objectives, and ensure the organization remains competitive in a rapidly changing digital landscape.

Conclusion

The CRM CoE leader is pivotal in aligning CRM initiatives with business strategies, fostering executive engagement, and dismantling silos. They manage change, nurture talent, ensure compliance, and champion customer centricity. Developing strategic relationship management with platform providers and optimizing CRM resources are crucial for achieving operational efficiency and customer engagement. The CoE leader's role ensures that CRM technology is a strategic asset driving the organization forward, leading to better integration, effective operations, and a stronger focus on customer satisfaction. A commitment to continuous adaptation and improvement will position the organization at the forefront of excellence in customer relationship management.

CHAPTER 7
CRM CoE Maturity

Executive summary

Chapter 7 explores the CRM Center of Excellence's growth journey, focusing on its evolution from reactive tasks to pro-active strategic design. This development is crucial for aligning CRM with the organization's core mission and strategic goals. Central to this alignment is the CoE's responsibility and accountability for carrying out the center's core Six Pillars. In practice, they are tailored to needs of individual organizations and, as such, form the backbone of each CoE's journey toward maturity. In general terms, the pillars call for the CoE to:

1. **Drive strategic alignment.**
2. **Establish governance.**
3. **Measure ROI realization.**
4. **Incorporate collaboration and interoperability.**
5. **Hone strategy stewardship and serve as an innovation hub.**
6. **Nurture acquisition of skills and expertise sharing.**

The chapter provides a practical guide to these roles through a recommended operating model for CRM projects, covering pre-delivery, delivery, and post-delivery phases. It underlines the importance of staffing, authority, and financial considerations in enhancing the CoE's effectiveness. By systematically addressing these areas, the CoE ensures continuous alignment with strategic objectives, driving growth, improving customer experiences, and fostering innovation.

The chapter highlights a structured approach to CRM CoE evolution designed to ensure the CoE remains agile and responsive to changing business needs. This approach, which includes setting clear goals, implementing best practices, and regularly reviewing and adjusting strategies based on performance metrics, provides a solid foundation for the CoE's success and its ability to drive organizational growth and competitive advantage.

Central to this journey is the endgame of customer centricity. The CoE must prioritize understanding and addressing customer needs and desires, both internal and external, to deliver exceptional value and build long-term relationships. This customer-centric approach benefits customers, strengthens the business, and sets it apart from competitors. By focusing on the goal of customer centricity, the CoE can make informed decisions that will fortify its overall effectiveness and transform it into a mature contributor to a culture of success.

CoE Maturity Journey

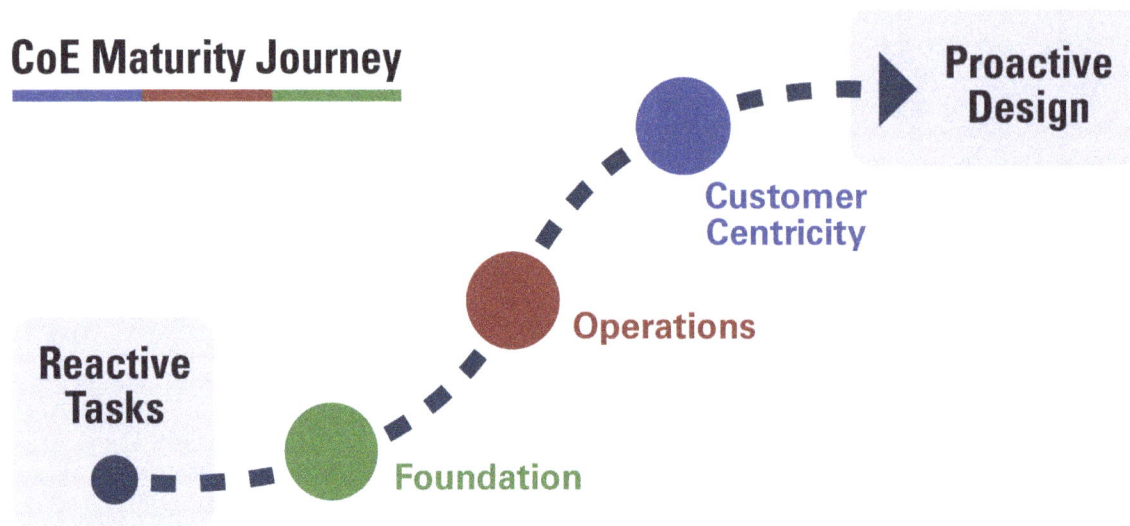

Figure 43: CoE's journey to maturity—from reactive tasks to proactive design.

This chapter follows the CRM Center of Excellence journey, exploring how this kind of CoE evolves and matures within an organization. Central to this lifecycle are the CoE's Six Pillars of responsibility and accountability. Every organization will manifest these roles differently, influenced by its unique operational dynamics and strategic objectives. The Six Pillars are:

1) Drive strategic alignment.
2) Establish governance.
3) Measure ROI realization.
4) Incorporate collaboration and interoperability.
5) Hone strategy stewardship and serve as an innovation hub.
6) Nurture acquisition of skills and expertise sharing.

As we progress through this chapter, we delve into the practical application of these pillars as the CoE's. We explore a recommended operating model and how the CoE effectively navigates its responsibilities in the pre-delivery, delivery, and post-delivery phases of CRM projects. Each phase presents unique challenges and opportunities, and we highlight the CoE's distinct responsibilities within each phase.

Strategic alignment and governance are foundational to the CoE, both of which ensure that CRM initiatives align with the organization's mission and adhere to established policies. This alignment facilitates ROI realization by maximizing the value derived from CRM investments. Collaboration and interoperability are crucial during the delivery phase, fostering cross-functional teamwork and seamless integration with other business systems.

The CoE also acts as a strategy steward and innovation hub, continuously assessing and refining CRM strategies to drive innovation and business growth. Management of skills and expertise ensures that the CoE team remains proficient and adaptable, capable of meeting evolving business needs. Throughout the chapter, we emphasize the importance of staffing, authority, and financial considerations, which significantly impact the CoE's effectiveness in fulfilling these roles.

The development focus of the CRM CoE pillars

Establishing a Center of Excellence (CoE) marks a transformative shift from reactive tasks to proactive organizational design. This transformation entails a changing emphasis on the three domains of foundation, operations, and customer centricity. In addition, the journey involves a progression in how the CoE focuses on developing each of the six work pillars as they are reflected in the core domains:

Foundation: The journey towards establishing a CRM CoE begins with a strong emphasis on this domain, which lays the groundwork for security, compliance, and governance. The initial efforts include establishing continuous integration and CI/CD deployment practices, as well as developing robust backup and recovery strategies. This foundational focus ensures the CRM platform is secure, compliant, and prepared to support scalable, efficient operations that are strategically aligned with organizational goals.

Operations: As the CoE evolves, the emphasis shifts towards the operations domain. This shift is crucial for ROI realization since it aims to maximize the benefits of CRM investments by improving the platform's functionality and usability. Optimizing these operational aspects allows the CoE to drive more efficiency and better integrate CRM strategies with broader business objectives, addressing the strategic alignment role.

Customer centricity: In the mature stages of the CoE's development, the focus increasingly turns toward the customer centricity domain. Initially, this domain centers on internal customers—the employees and teams using the CRM system—aiming to enhance their adoption rates, satisfaction, and productivity. This focus supports the roles of 1) collaboration and interoperability and 2) skill and expertise management. Improving the internal user experience paves the way for a deeper focus on end customers, aiming to enrich their interactions and engagement with the organization. By adopting a customer-centric approach, the CoE supports the strategy steward and innovation hub roles and drives business growth through enhanced customer loyalty and stronger relationships, positively influencing ROI realization. AI technologies can accelerate the development of CRM CoEs by providing advanced analytics, automating processes, and enhancing customer engagement.

The Six Pillars, the backbone of the CoE's development, are central to its work. These roles are designed to integrate the what and why of the CoE, ensuring that CRM practices are aligned with the organization's strategic goals and facilitating value realization from CRM investments. Each role is tailored to address the unique needs and dynamics of the organization, underpinning a structured approach to CRM excellence.

The CRM CoE's path across pillars and domains

1. Drive strategic alignment

Establishing a Center of Excellence enables a company to utilize the customer relationship management platform as a strategic asset, crucial for the company's vision and long-term success. A robust vision and clear purpose are essential when launching a CRM CoE. This begins with thoroughly understanding the company's CRM history, its organizational structure, and the initial reasons behind the platform's adoption. Understanding today's leadership objectives with the CRM platform is equally crucial.

Often, companies implement CRM systems without initially recognizing the need for a CoE. Aligning the CoE vision and purpose solely with the tactical goals of the initial CRM implementation can lead to challenges, since the initial implementation leaders often encounter significant tactical delivery pressures that may conflict with best practice processes and the broader vision and purpose of the CoE.

Effective CoEs delineate their vision, purpose, and functions from the delivery component. The CoE focuses on standards, best practices, security, compliance, strategy, operations, and administration, while CRM delivery leadership optimizes delivery processes in the CoE's framework. This separation helps maintain the platform's integrity and prevents operational pressures from causing deviations from established policies.

Crafting and communicating a vision that aligns with the company's core values and objectives makes the CoE's vision a guiding force for its future and fosters meaningful dialogue. Effective communication of this vision enhances understanding within the CoE, translates the vision into actionable outcomes, and keeps all stakeholders informed. When presenting the vision, it should be conveyed passionately and creatively to inspire a team that, as a result, feels personally invested.

A primary role of the CoE involves setting standards, best practices, security, compliance, strategy, operations, and administration for the CRM platform, ensuring alignment with broader company goals. Laying this groundwork in the foundation domain prevents deviations due to operational pressures and provides a robust base for advancing into the operations and customer centricity domains.

2. Establish governance and a CoE operating model

Chapter 2 explored the standards and best practices essential for establishing governance across the CRM Center of Excellence functions. This section focuses on the governance structure, examining how the CoE is strategically organized within the company relative to other units. The evolution of a CoE from reactive tasks to proactive organizational design is characterized by a shifting emphasis across the three domains, which inform the choice of operating model.

CRM CoE operating models are multifaceted and structured along **four critical dimensions—delivery, control, placement, and program versus product.** The choice of placement along these dimensions significantly influences the CoE's effectiveness and alignment with corporate strategy. Dimensions include:

Delivery. The first dimension, delivery, is where the CoE's role is pivotal. It can function as either a governor or an implementer. As a governor, the CoE assumes the role of a policymaker, setting standards and best practices. While not directly managing daily CRM operations, this function aligns with the foundation domain to establish robust governance. On the other hand, as an implementer, the CoE takes an active role in both 1) setting policies and executing CRM initiatives and 2) managing CRM systems' deployment and ongoing management. A mature CoE will not be both governor and implementer for delivery. It will sit in the governing seat. Otherwise, it will always be in conflict with itself over which role to prioritize, and neither will get its full due.

Control. The second dimension, control, offers varied models, from centralized or decentralized, to a hub-and-spoke collaborative model. Central control is a governance model where all decisions and CRM management practices are consolidated at a single point—a typical approach in the early stages of the foundation domain. Decentralized control, more in line with the customer centricity domain, distributes decision-making across various departments to enhance responsiveness and local relevance. The hub-and-spoke model, a hybrid of central oversight and decentralized execution, is ideal for balancing strategic alignment with operational flexibility. These models offer different decision-making approaches, each with benefits and implications for the CoE's operations. At the early stages of the CoE's journey, centralized control may make more sense, but developing a more balanced, more mature approach is essential.

Placement. The third dimension is placement within the company, either in IT or business operations. A emphasis on IT positions the CoE within the IT department, emphasizing technological governance and system integrity, which are crucial in the foundation domain. Business-centric placement integrates the CoE within business units, focusing on CRM's role in driving business objectives and supporting the customer centric domain. In most companies, straddling IT and business does not suit the organizational structure, and generally, the CoE will be placed in one or the other. If a company's structure allows for blended control, it may be worth exploring what that looks like. Ultimately, even when blended control is desired, there will be, in practice, a single point of control.

Program versus product. The fourth dimension contrasts a program approach with a product approach. The program approach treats CRM initiatives as discrete projects with specific timelines, often used in the initial operational setup. The product approach views CRM as a continuously developing and refining entity crucial for sustained customer-centric innovation.

Implementing the CoE operating model

Implementation of the CoE operating model involves several key steps. First, it requires establishing a structure to determine the focus on governance versus implementation, the locus of control, and the CoE's place in the hierarchy. It also requires a basic decision about whether to adopt a product or program approach.

Next, clearly defining roles and responsibilities within the CoE is essential to ensure effective management and operational clarity. Setting up communication channels is critical for facilitating effective communication with other parts of the organization and guaranteeing integrated CRM management. Continuous monitoring and optimization are necessary, as well as regularly adjusting the operating model to align with the CoE's stage of maturity and the company's strategic objectives.

By choosing an appropriate operating model, the Center of Excellence enhances its ability to support the organization's strategic objectives and effectively reflect its maturing role. This alignment is crucial for maximizing the CoE's effectiveness, driving sustained business value, and supporting the CoE's journey from foundational governance to operational excellence and customer-centric innovation.

CRM CoE Operating Models

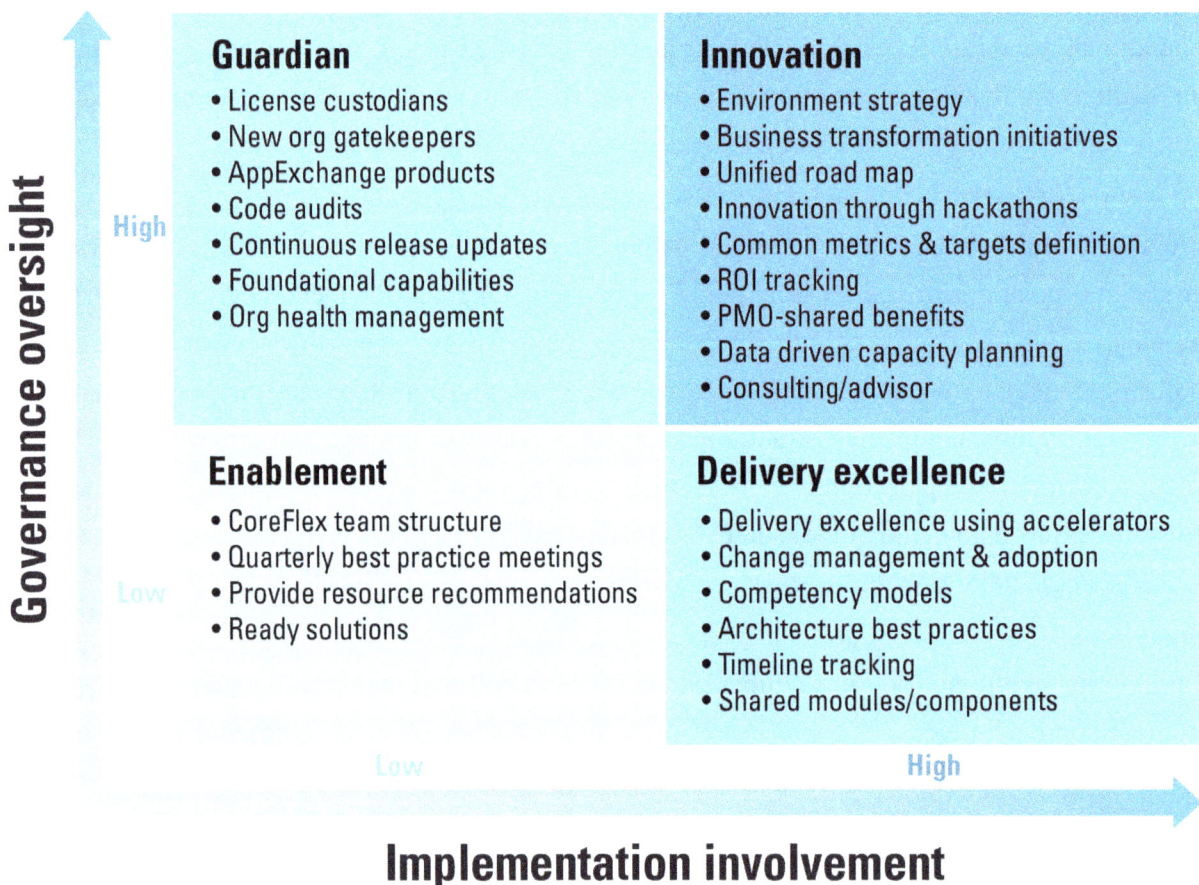

Governance oversight (vertical axis, High to Low)
Implementation involvement (horizontal axis, Low to High)

Guardian
- License custodians
- New org gatekeepers
- AppExchange products
- Code audits
- Continuous release updates
- Foundational capabilities
- Org health management

Innovation
- Environment strategy
- Business transformation initiatives
- Unified road map
- Innovation through hackathons
- Common metrics & targets definition
- ROI tracking
- PMO-shared benefits
- Data driven capacity planning
- Consulting/advisor

Enablement
- CoreFlex team structure
- Quarterly best practice meetings
- Provide resource recommendations
- Ready solutions

Delivery excellence
- Delivery excellence using accelerators
- Change management & adoption
- Competency models
- Architecture best practices
- Timeline tracking
- Shared modules/components

Figure 44: Salesforce's CoE operating models.

Delivery: Governor vs. implementor

The CoE operating delivery model varies in its defined responsibilities regarding implementation (delivery) versus governance. Many CRM CoEs have part or full responsibility for delivery, which can undermine the CoE's purpose, as delivery deadlines often overshadow compliance and security requirements. These operational models can be represented in a grid with four primary quadrants, each reflecting a unique combination of governance and implementation responsibilities (Figure 44).

In the Enablement quadrant, characterized by low governance oversight and low implementation involvement, the CoE functions as a powerful enabler. It provides resources, best practices, and recommendations without direct involvement in project delivery, empowering the organization to develop self-service capabilities through flex team structures, quarterly best practices meetings, resource recommendations, and ready solutions.

The Innovation quadrant, with low governance oversight and high implementation involvement, features a high level of strategic focus. This model is less about strict governance and more about leading and executing transformative projects, fostering innovation through hackathons, and tracking ROI and data-driven capacity planning. It's an exciting space where the CoE is at the forefront of transformational initiatives.

In the Guardian quadrant, characterized by high governance oversight and low implementation involvement, the CoE primarily focuses on governance. It ensures compliance and security without directly driving project delivery. It handles tasks such as license and vendor contract procurement and management, new CRM instance gatekeeping, AppExchange product approvals, code audits, continuous release updates, foundation capabilities maintenance, and CRM instance health management.

The Delivery Excellence quadrant, with low governance oversight and high implementation involvement, emphasizes efficient and high-quality delivery mechanisms with less focus on overarching governance. This model supports using accelerators to enhance delivery, change management, adoption strategies, development of competency models, implementation of architecture best practices, timeline tracking, and sharing modules and components.

Control: Decision-making structure

The control and decision-making structure is a strategic decision that can significantly impact the effectiveness of CRM initiatives across an organization. The selection between centralized, decentralized, and hybrid models, each with unique benefits and challenges, is a crucial task that requires careful consideration and understanding of the organization's needs and goals.

A centralized model acts as the single point of control and decision-making, often used in smaller organizations or those with centralized IT structures. This model ensures consistency in processes and standards, aligning CRM initiatives with organizational goals. However, the model may need more flexibility and responsiveness to meet individual department needs.

In contrast, a decentralized model allows individual business units or departments to operate their own CRM teams independently. This model offers greater autonomy and specialization, enabling rapid adaptation and innovation aligned with customer demands. The drawbacks include potential inconsistencies in standards and practices, challenges in optimizing talent allocation, and difficulties maintaining a unified customer experience.

Hybrid models blend the benefits of centralized and decentralized approaches, providing a balanced approach to CRM management. The hub-and-spoke model features a central CoE (hub) that sets overall strategy, governance, and standards, while decentralized teams (spokes) adapt these to their specific needs. This model balances consistency with flexibility, making it practical for large, complex organizations. Implementation requires robust communication and a strong governance framework to ensure cohesive efforts and optimal resource allocation.

A center-led model, like the hub-and-spoke, provides central guidance and best practices but involves more collaborative decision-making with business units. This model enhances cooperation and ensures alignment between the CoE and various departments.

When selecting a control and decision-making structure, factors such as the organization's size, complexity, customer base, and strategic objectives must be considered. These practical touchstones can help determine the most suitable model to enhance CRM effectiveness and drive customer-centric growth in a specific organizational context.

Implementing a hub-and-spoke model involves several critical steps and considerations. The central hub manages the overall CRM strategy, ensuring data integrity and overseeing vendor relationships. It also serves as the repository of best practices and innovations. Decentralized spokes implement and customize CRM strategies to fit their specific needs, maintaining autonomy within the overarching framework set by the hub. Regular meetings and shared platforms are required to maintain a cohesive plan. At the same time, flexibility and scalability allow the system to adapt as the organization grows, integrating new spokes without disrupting the CRM ecosystem. This model supports a consistent organizational strategy and caters to localized needs, thus enhancing the overall customer experience. Successful implementation requires careful planning, a clear definition of roles and responsibilities, and a commitment to ongoing collaboration and communication.

Placement: IT or business

The strategic placement of a CRM CoE within an organization is crucial for maximizing its impact on customer engagement and achieving business outcomes. Whether to align the CoE with the IT department or business units significantly influences its ability to drive effective CRM strategies.

Aligning the CoE with IT provides deep technical expertise and infrastructure knowledge, which is crucial for maintaining a robust and scalable CRM system. This alignment ensures the technological sophistication of CRM initiatives. However, there is the risk of a disconnect between the IT-focused CoE and the organization's immediate business goals. The technical orientation may need to fully align with the need for customer-centric outcomes, necessitating measures to bridge this gap.

On the other hand, aligning the CoE with business units places it closer to strategic business objectives and decision-makers. This proximity facilitates initiatives that directly impact customer engagement and sales, helping to align CRM strategies with business needs. However, this approach may expose gaps in technical expertise, which requires close collaboration with IT to ensure the technological robustness of CRM solutions.

Several strategies can be employed to effectively manage the alignment of the CoE, whether with IT or business. Forming cross-functional teams that include members from both IT and business units fosters a shared understanding and collaborative environment. Regular joint strategy sessions help synchronize the efforts of IT and business, ensuring that CRM initiatives are both technically sound and aligned with business goals. Councils are pivotal in supporting the CoE's strategic alignment, operational efficiency, and delivery excellence. These councils facilitate communication and decision-making between IT and business, clarify the roles and responsibilities of each council, and act as a bridge, reinforcing the importance of collaboration to address emerging challenges effectively.

Regardless of the CoE's placement, the cornerstone of its success lies in effective communication, mutual understanding, and shared goals between the technology and business sides of the organization. Emphasizing collaboration can mitigate the potential challenges of either alignment, ensuring that the CoE operates effectively and drives substantial business outcomes.

Program vs. product approach

The shift from a program approach to a product approach in implementing CRM systems marks a significant evolution in how organizations manage and perceive their CRM strategies. This change aligns more closely with modern business needs for agility, continuous improvement, and customer centricity.

For a deeper understanding of this transition, see *The Product Mindset: Succeed in the Digital Economy by Changing the Way Your Organization Thinks*, by David H. DeWolf and Jessica S. Hall.[39] It is a highly recommended resource that explores the underlying principles and benefits of adopting a product mindset in business operations.

Here's how this shift impacts various aspects of CRM implementation. The program approach, focusing on achieving specific business objectives within a defined time frame and budget, may seem rigid in a rapidly changing business landscape. In contrast, with its adaptability and continuous evolution, the product approach offers a more reassuring strategy. It treats the CRM system as a living entity, always ready to adapt to changing business needs and user feedback. It focuses on long-term engagement and incremental improvement and emphasizes ongoing value creation.

In terms of methodology, the program approach often utilizes a traditional waterfall project management methodology, following a linear, step-by-step process structured around completing specific projects or milestones. In contrast, the product approach leverages Agile methodologies, prioritizing flexibility, quick iterations, and responsiveness. This enables ongoing adaptations to the CRM system based on real-time feedback and evolving business requirements.

Regarding goals, the program approach strives to deliver the CRM system within pre-set scopes, timelines, and budget constraints, measuring success by adhering to these constraints and strategic objectives. The product approach, on the other hand, is all about continuous value creation. It continuously meets user needs and enhances customer satisfaction, driving business growth. Success is evaluated based on the system's adaptability and sustained value to the organization and its customers, instilling confidence in its long-term benefits.

Regarding budgeting and funding, the program approach often features a fixed budget allocated based on estimated costs to complete defined projects, with little room for adjustment once set. The product approach, however, allows for more dynamic and flexible budgeting, enabling funds to be reallocated to address changing priorities or new opportunities as they arise.

Management and governance under the program approach are typically handled by a project management office (PMO), which focuses on project coordination and adherence to budgets and timelines. In contrast, a product manager with a cross-functional team usually oversees the product approach, facilitating quicker decision-making and adjustments based on ongoing product assessments and market feedback.

The outcomes of these approaches also differ. Once the CRM system is delivered using the program approach, it might meet its initial objectives but could require significant updates or a new program for

---

39   David H. DeWolf and Jessica S. Hall, *The Product Mindset: Succeed in the Digital Economy by Changing the Way Your Organization Thinks* (New York: HarperCollins Leadership, 2019).

future enhancements. The product approach results in a CRM system that is continually updated and refined, allowing for quick adaptation to new challenges or opportunities. The system evolves as a so-called "living product."

Adoption and change management in the program approach includes a distinct phase focused on training and transitioning users to the new system toward the end of the project. The product approach incorporates ongoing change management with continuous user engagement and feedback integration, ensuring the system remains relevant and practical.

Implementing the product approach involves creating a dynamic road map for the CRM system's evolution post rollout. This road map aligns system enhancements—like analytics improvements, UI/UX upgrades, and integration projects—with overarching business objectives. By treating these enhancements as part of a continuous development cycle rather than isolated projects, the organization gains a sense of empowerment in managing its CRM system. This strategy optimizes resource allocation and maintains the CRM's relevance in a competitive business environment, giving the audience a strategic edge.

In summary, the move towards a product approach in CRM implementations offers a more flexible and sustainable framework that aligns with modern business dynamics and the need for continuous innovation. This approach helps organizations stay responsive to customer demands and market changes, ensuring their CRM systems grow and evolve with their strategic needs. For those interested in deepening their understanding of this methodology, reading *The Product Mindset* can provide valuable insights into how changing an organization's thinking can lead to success in the digital economy.

Pre-delivery, delivery, post-delivery stages

The CRM Center of Excellence role in guiding the lifecycle of a CRM system is influenced by its maturity level and the chosen operating model. The choices range from **delivery** (governor versus implementer) and **control** (centralized versus decentralized versus hub-and-spoke) to **placement** within the company (IT versus business) and **approach** (program versus product). The choice of model significantly impacts how the CoE operates across the CRM system's pre-delivery, delivery, and post-delivery stages.

Pre-delivery: At this point, strategic planning and governance are essential. The maturity of the CoE and its positioning, whether IT or business-aligned, dictate its capacity to establish comprehensive governance and strategic alignment. For instance, a mature, business-aligned CoE may excel in aligning CRM strategies with business goals, while an IT-aligned CoE might focus more on technical standards and infrastructure. Choosing between a centralized or hub-and-spoke model affects how requirements are gathered and analyzed. Centralized models might streamline this process by consolidating requirements across the organization, whereas hub-and-spoke models allow for tailored requirements that cater to specific business unit needs. The criteria for selecting technologies and vendors can vary significantly depending on whether the CoE operates under a program or product approach. A product approach, for example, may prioritize scalability and integration capabilities to support continuous improvement.

Delivery: Project management and oversight are influenced by whether the CoE adopts a governor or implementer model. A governor model in a mature CoE may focus more on setting standards and overseeing compliance. In contrast, an implementer model would take a hands-on approach to managing the CRM implementation. The choice of control also influences project management, with centralized models potentially providing stricter oversight compared to decentralized models.

The placement of the CoE, whether IT or business-centric, impacts how the CRM is configured and customized. An IT-centric CoE might emphasize technical aspects such as system architecture, whereas a business-centric CoE would focus on aligning configurations with business processes. The approach, whether focused on program or product, affects how training and change management are conducted. A program approach might see these as finite, project-bound activities, whereas a product approach integrates them as ongoing processes that evolve with user feedback and system upgrades.

Post-delivery: At this stage, performance monitoring and analytics are key. The maturity level of the CoE is reflected in its ability to monitor performance and utilize analytics effectively. More mature CoEs will likely have sophisticated tools and processes for continuous evaluation and improvement. The operating model dimension of control, whether centralized, decentralized, or hub-and-spoke, dictates how the CRM system evolves post-delivery. Centralized models may focus on standardization across the organization, while decentralized models allow individual units to innovate and adapt independently.

These dynamics underscore how the selected operating model and the maturity of the CoE shape each phase of the CRM system's lifecycle. Effective management of these models ensures that the CRM CoE can adapt its strategies and operations to meet current needs and prepare for future challenges, thus maximizing the overall impact of the CRM initiatives across the organization.

3. Measure ROI realization

Chapter 4 focused on ROI justifications by CRM CoE function. Here, we pick up the conversation about crucial CoE metrics and budgets.

Incorporating the five CRM CoE Cultural Aspects—data, people, process, technology, and agility—as lenses for ROI and weaving in the previously described metrics, we can more comprehensively understand and measure the impact of CRM initiatives. Here's how each aspect can drive specific metrics, enhancing ROI realization through a nuanced understanding of organizational culture.

Data: Data-driven decision-making ensures CRM initiatives are aligned with insightful, accurate data, enhancing CRM effectiveness. In a centralized funding model, data integrity and security are prioritized, which supports data quality and accuracy—crucial metrics for ROI.

Metrics: Data quality, data accuracy

People: Fostering a culture of continuous improvement and high performance ensures the workforce can implement strategic visions that enhance customer relationships. This alignment can be measured through NPS and CSAT, with centralized funding ensuring uniform organizational training and development initiatives.

Metrics: Net Promoter Score (NPS), Customer Satisfaction (CSAT)

Process: Streamlining processes to align with strategic goals enhances operational efficiency. Centralized funding can enforce consistent process improvements across the organization, reducing cycle times and improving response rates, directly impacting the ROI

Metrics: Cycle times, response rates

Technology: Strategic technology deployment under a centralized model ensures that investments are made in technologies that offer the best value and are aligned with business strategies. This can be measured through system uptime and system performance rates.

Metrics: System uptime, system performance

Agility: Agility allows the organization to swiftly adjust strategies in response to changing business conditions. This aspect is crucial in centralized and distributed models but needs careful management to ensure that agility does not compromise strategic alignment. On time and on budget delivery rates are vital metrics here.

Metrics: On time delivery, on budget delivery

Cross-functional coordination and impact on ROI

Effective cross-functional coordination is required to maximize the ROI from CRM initiatives and ensure that all departments work toward unified CRM goals, regardless of the funding model. Customer metrics such as NPS, CSAT, CLV, and churn rate reflect the direct impact of CRM initiatives on customer engagement and satisfaction. High scores in these areas indicate effective cross-departmental collaboration and alignment. Financial metrics, including cost savings and revenue growth, are directly influenced by how well CRM strategies are integrated across the organization. Effective cross-functional coordination ensures that CRM implementations in general contribute to these financial benefits.

Process metrics, such as cycle times and response rates, improve through streamlined processes resulting from effective cross-functional coordination, enhancing overall operational efficiency. Technology metrics, like adoption rates and system uptime, indicate CRM systems' successful integration and functionality, which are essential for maintaining continuous business operations. Project metrics, including on time and on budget delivery, reflect the effectiveness of project management and planning capabilities across functions, which are crucial for maintaining budget discipline and meeting timelines.

Managing the intricacies of budgets and relationships determines how CRM initiatives are prioritized, funded, and executed, impacting their overall success and return on investment. In a centralized funding model, a central authority, usually the CoE, controls and allocates all financial resources for CRM initiatives. This centralized approach allows for a strategic overview of investments and streamlined resource allocation, ensuring that projects align with the organization's primary objectives. It simplifies the approval process and maintains consistency in CRM standards across the enterprise. However, this model requires careful management to avoid perceptions of bias and to ensure that specific departmental needs are addressed.

In contrast, distributed funding empowers individual departments to manage their budgets for CRM activities. This model fosters direct accountability and motivates departments to implement CRM solutions that cater to their operational needs, potentially driving innovation and customized solutions. However, without stringent guidelines, this approach might lead to redundant efforts or projects that do not align well with the overarching CRM strategy.

Whether the funding is centralized or distributed, cross-functional coordination remains a cornerstone of achieving effective CRM outcomes. This coordination involves building and maintaining robust relationships and open communication channels across various organizational domains, including IT, marketing, sales, customer service, and finance. Effective coordination practices establish clear communication channels that facilitate regular updates and feedback across departments, conduct regular stakeholder meetings to ensure all parties are aligned and informed, and utilize collaborative platforms to share insights, progress, and data among team members. This emphasis on open communication makes a team feel included and valued, fostering a sense of unity and shared responsibility.

The CoE, acting as a central hub, coordinates these efforts and mediates between varying departmental interests. This ensures that each department contributes to a cohesive and unified CRM strategy. By aligning the CRM CoE Cultural Aspects with strategic budgeting approaches and cross-functional coordination, the organization can ensure that CRM initiatives meet immediate goals and contribute to long-term business growth. This comprehensive approach to measuring ROI highlights the interconnectedness of Cultural Aspects, financial management, and operational execution, providing a robust framework for assessing the success and impact of CRM strategies. This reassures a team that there is a structured process in place, instilling confidence in the CRM strategy.

4. Incorporate collaboration and interoperability

In this chapter, we have intricately woven collaboration and interoperability into the narrative of the CRM Center of Excellence's journey. These factors merit focused examination as distinct yet interconnected elements essential to the CoE's ongoing success and evolution. Collaboration and interoperability extend beyond technology integration, encompassing data, cross-functional skill development, and process harmonization across the organization.

As the CoE transitions from its initial setup phase into a phase of continuous enhancement, several strategic projects emerge to solidify the CoE's role as an innovation hub and ensure it remains responsive to the organization's changing needs. These factors are critical for leveraging the CRM system's full potential.

Efforts to integrate across business systems aims to streamline workflows and enhance data accuracy across systems. This ensures seamless data flow and operational efficiency, reducing errors and information silos. Advanced analytics and AI implementation extract deeper insights from vast customer data pools, enabling predictive analytics and personalized customer engagement strategies, transforming data into actionable intelligence. Enhancing mobile and web interfaces improves the user experience for CRM system users on all platforms, leading to increased user engagement and satisfaction, which is crucial for higher adoption rates and productivity.

Ongoing training sessions and real-time in-app training equip users with the necessary skills and knowledge to utilize the CRM system fully. This enhances user competence and confidence, making CRM capabilities more effective. Regular CRM reviews and updates align the CRM system with new market trends and customer expectations, ensuring that CRM remains a relevant and powerful tool for customer engagement and business strategy.

By focusing on these factors, the CoE deepens its integration within the business's operational fabric and highlights its commitment to fostering a culture of continuous improvement and adaptability. This proactive approach to managing collaboration and interoperability is essential for the CoE to continue delivering significant business value, enhancing customer relationships, and driving sustained organizational growth.

5. Hone strategy stewardship and serve as an innovation hub

In its dual capacity as strategy steward and innovation hub, the CRM Center of Excellence navigates a delicate balance between safeguarding established strategies and pioneering new pathways for growth. The integration of AI into CRM strategies necessitates vigilant awareness of this balance. The allure of new tools can lead to hasty adoption for some and prolonged reluctance for others. As the CoE matures, this balancing role involves continuous assessment and recalibration of strategic directions influenced by key performance metrics.

As a strategy steward, the CoE is responsible for maintaining strategic consistency across the organization, ensuring compliance with internal policies and external regulations, and guiding long-term strategic alignment with the organization's overarching business goals. As an innovation hub, the CoE explores emerging technologies to enhance CRM effectiveness, fosters a culture of creativity and experimentation to improve customer interactions and operational processes, and pilots innovative solutions that could redefine business processes and customer engagement.

The balance between maintaining strategic guidelines and fostering innovation is not static. It shifts with the CoE's maturity, offering exciting opportunities for growth and innovation. Initially, the focus may be more on establishing robust governance, whereas later stages increasingly prioritize innovation, utilizing metrics to guide these efforts.

CoE leaders must choose key performance indicators (KPIs) that reflect leadership interests, recognizing that these preferences will evolve with the CoE's maturity. These metrics are not just numbers; they are the compass that guides our strategy stewardship and innovation hub roles. Key reporting areas and metrics include project and product progress, user adoption and engagement, business impact, system performance and integration, data quality and management, feedback, challenges and risks, and success stories and lessons learned.

Additional KPIs include delivery metrics such as deployment frequency, mean lead time for changes, mean time to recovery, and change failure rate, which measure operational efficiency and system resilience. Governance KPIs like CRM instance health, access controls, data growth, and audit compliance uphold system integrity and compliance. Operational KPIs for user, release, environment, license management, and tech debt are vital for continuous system management and innovation. Business impact KPIs, including revenue, CSAT, and NPS, directly measure the CRM's contribution to business growth and the strength of customer relationship.

Effective management of these responsibilities and dynamic adjustment of KPIs ensure that the CoE is both a guardian of strategy and a catalyst for innovation. This dual role allows maintaining robust CRM strategies aligned with business objectives and adaptable to new opportunities, driving sustained organizational growth and competitive advantage.

6. Nurture acquisition of skills and expertise sharing

Understanding the evolution of staffing and authority within a CRM Center of Excellence (CoE) matters, as the organization matures. The transition from initial setup to a mature CoE affects how roles and skills are structured and utilized, impacting the CoE's strategic deployment and operational success.

As the CoE evolves, its roles must adapt to shifting needs and expanded capabilities. In the early stages, dedicated roles might focus on foundational tasks such as setting up CRM systems and establishing basic functionalities. As the CoE matures, these roles evolve to focus on advanced development, where developers shift from basic customizations to integrating complex functionalities involving AI and advanced analytics. Architectural innovation also becomes critical, with architects moving from establishing a basic framework to optimizing the CRM for scalability and advanced security to support growing organizational needs. CRM administration roles transition from general support to managing increasingly complex system integrations and data management tasks. These dedicated roles allow for deep specialization and can drive significant advancements in CRM capabilities as the CoE matures.

In a matrixed model, the integration of CRM tasks with other business functions becomes more pronounced as the CoE matures. This model supports cross-functional expertise, with roles such as business analysts and product managers integrating CRM strategy more deeply with enterprise-wide digital transformation initiatives. Dynamic resource allocation, as projects become more complex and involve more stakeholders, demands the strategic assignment of experts from various domains to CRM projects. Developers in a matrixed environment engage in more strategic projects, influencing CRM and broader IT initiatives, and require refined time management and prioritization skills.

As the CoE matures, the operational focus shifts, requiring adjustments in role definitions and responsibilities. The early stage emphasizes basic system setup, user training, and initial integration, with roles focused on implementation and foundational practices. In the growth stage, the focus shifts to system optimization, scaling user adoption, and integrating advanced technologies, with roles evolving to manage more complex workflows and data systems. In the maturity stage, the strategic refinement of CRM capabilities drives innovation and business transformation, requiring much more expertise in data analysis, customer experience design, and change management.

Underscoring the role of strategic staffing in a mature CoE, we see that it involves continuous learning and development to ensure team members are equipped with the latest CRM technologies and methodologies. Role flexibility is also essential, allowing roles to evolve in response to new challenges and opportunities and ensuring staffing aligns with current and future needs. Strategic staffing aligns human resources with long-term business goals and the evolving technological landscape, ensuring the CoE remains a robust driver of business value.

A customer-centric CRM Center of Excellence

Before concluding this chapter, we want to return to the beginning and our discussion of the endgame in CoE maturity: the customer centricity domain. We said that the journey toward maturity moved the CoE emphasis from reactive tasks to a proactive strategy, with the focus moving from the foundation domain to the operating domain and then the customer centricity domain. Here, we aim to highlight the context for a customer-centric CRM Center of Excellence.

To create a customer-centric CRM CoE, necessitates comprehension of our customers and their needs. Internal customers include various teams such as those in strategy, information security, product ownership, developers, business analysts, testers, executive leadership, senior leadership, finance, procurement, legal, products, operations, sales, marketing, customer service, business continuity, audit and compliance, and other stakeholders. External customers include end customers, partners, vendors, channel partners or resellers, community organizations, and other stakeholders.

Customers seek efficiency, reliability, innovation, and value in our solutions. They require seamless integration, user-friendly interfaces, robust support, continuous improvements, quick response times, transparency, and effective communication. They dislike complexity, inefficiency, poor support, and lack of transparency.

Customer feedback is a valuable resource in shaping our CRM CoE strategies. We gather this feedback through surveys, feedback forms, interviews, focus groups, analytics, and usage data. This information is not just data but a catalyst for change. It is transformed into actionable strategies, leading to continuous improvement, enhanced communication, and comprehensive training and support. By actively seeking and acting on customer input, we can ensure our strategies are always customer-centric.

Adopting a customer-centric approach is a commitment to our customers. It involves regular engagement with all kinds of customer segments, prioritizing critical needs, actively anticipating future needs, measuring customer satisfaction, and fostering collaboration between internal teams and external partners. By deeply understanding our customers and acting on their feedback, we ensure that our CoE delivers exceptional value and builds long-term relationships. This approach strengthens our business and sets us apart from our competitors. Knowing the endgame is customer-centric helps us make the appropriate choices along our route and more expeditiously make that journey.

Conclusion

The CRM CoE's maturity journey involves transitioning from foundational tasks to strategic initiatives. This evolution is characterized by a shifting focus across three domains: foundation, operations, and customer centricity. The development of the Six Pillars—1) drive strategic alignment 2) establish governance

3) measure ROI 4) incorporate collaboration and interoperability 5) hone strategy stewardship and an innovation hub, and 6) nurture acquisition of skills and expertise sharing—guides this journey.

Critical stages of CoE maturity include:

1. Foundation domain: Establishing governance, compliance, and security.
2. Operations domain: Enhancing system functionality and user adoption.
3. Customer centricity domain: Focusing on internal and external customer satisfaction.

The CoE adapts its structure and strategy to meet the evolving needs of the business, from setting governance frameworks to driving innovation. Whether centralized or decentralized, dedicated or matrixed, operational models significantly influence the CoE's effectiveness across CRM lifecycle stages.

Metrics and KPIs tailored to the CoE's evolving stages provide a data-driven approach to measure success and inform strategic decisions. This ensures that CRM initiatives align with broader organizational goals.

Ultimately, the CoE's success hinges on its ability to integrate seamlessly with the organization's culture, adapting strategies and operations to meet current needs and anticipate future challenges. A robust, adaptive, forward-looking CoE drives long-term business value through effective CRM management, making it more likely that the organization will remain competitive and innovative.

A Tale of Decentralized Decision-Making

In a place far away and long ago, I ran a small lodging business. My decision-making rule was that there were "one-headed," "two-headed," and "three-headed" decisions. I expected employees to learn over time and act accordingly as to which of these would fit the decision they faced.

Did a customer need a discount because the guest's room was not prepped when they arrived? Only if we had nowhere else to put them, and it was late at night. Depending upon the circumstances, that might be a one- or two-headed job. In one-headed jobs, you decide based on your best judgment and go with it. In two-headed jobs, you consulted a colleague, and together, you decided. If no agreement was possible, it was a three-headed job. In three-headed jobs, you brought in a superior to help make the decision.

Most decisions at the task level are one-headed and two-headed jobs. As the complexity rises, they become two- and three-headed decisions. The roof is leaking into an upstairs unit. Do we move the guests to another unit and wait until Monday to call the roofer? It's a one- or two-headed job. But what if no other rooms are available, or if it's forecast to rain all week and the top floor is not the only damaged floor? The decisions become different, and they become three- and many-headed decisions with tasks for many.

This notion of control and empowerment will vary with industry and the organizational culture. What is clear is that controlling everything and making every decision is not helpful or efficient. If that ability to loosen control and delegate decision-making is scary, look back at the Cultural Aspect section of this book and the function comprising the Cultural Aspect—people. What skills and organizational safety nets (processes and technology) need to be in place so you can empower your people to do their best work and make their best decisions?

Charlie Havens

CHAPTER 8
Navigating the Enterprise

Executive summary

Chapter 8 explores the evolution of CRM Centers of Excellence from initial establishment to achieving broader optimization within the enterprise. Early-stage leaders focus on understanding business processes and gaining support, while mature leaders manage complex integrations, align CRM strategies with business objectives, and adapt to technological advancements.

Initially, leaders must influence without authority by building relationships and advocating for the benefits of customer relationship management. Over time, they evolve into strategic advisers who tend to guide company-wide strategy through CRM insights. Strategic alignment begins with quick wins and matures into shaping the company's overarching approach to support evolving business models and industry demands.

Change management starts with specific projects or departments, laying a foundation for broader CRM adoption. As the CoE matures, change management fosters a culture of continuous improvement and innovation to handle extensive transformations across the enterprise.

Accountability and responsibility in the early stages focus on CRM projects within specific departments. In mature stages, this expands to sustaining CRM system performance across the organization, coordinating with multiple business units and external partners.

Championing best practices progresses from establishing standards for data quality, user training, and system setups to leading industry innovations and setting benchmarks that enhance the internal CRM ecosystem and influence broader market practices.

Addressing seasonality peaks starts with basic analytics and forecasting, advancing to sophisticated predictive analytics and machine learning. This allows mature CoEs to optimize resources and improve customer engagement strategies across business cycles.

Building an internal CRM CoE movement begins with rallying early adopters and securing endorsements through project successes. As the CoE matures, this movement embeds CRM principles into the corporate culture through ongoing engagement, continuous education, annual CRM collaboration events, community building, and recognition programs.

Connecting with the external ecosystem starts with learning and benchmarking against best practices. As the CoE matures, leaders participate in and lead discussions in external ecosystems, contributing to thought leadership, collaborating on cross-industry initiatives, and influencing CRM technology developments.

The shift towards industry solutions by organizations like Salesforce and Microsoft Dynamics reflects a growing recognition that businesses need comprehensive solutions tailored to their specific industries. This trend reshapes the CRM landscape, positioning companies that embrace it for success in the digital age.

Ultimately, the CRM CoE Framework brings a common language and consistent methodologies to the organization. It ensures all departments align in understanding and executing CRM strategies, promoting a unified approach to achieving business objectives.

Navigating the CoE's evolving role

As CRM Centers of Excellence (CoEs) mature, the strategies and challenges of navigating the enterprise evolve. Initially focused on establishing credibility and aligning with strategic business goals, mature CoEs shift towards optimizing and expanding their impact both internally and externally. This evolution underscores the importance of adaptability, strategic foresight, and an increasing reliance on collaboration across departments and beyond the organization. The individual's role in this collaborative effort is crucial.

As CRM CoEs mature, their leaders are strategic visionaries. They are empowered to navigate complex landscapes, influence without authority, manage far-reaching change, and drive the organization toward sustained CRM excellence. Their shift in role is a testament to their capabilities and influence.

For example, as a strategic visionary within a company, can you find ways that AI might facilitate enterprise-wide collaboration across systems, ensuring seamless data flow and improved output across departments? This feasibility varies by company structure, CoE development, and the organization's tactical AI expertise and capabilities.

In the early stages, CoE leaders focus on understanding existing business processes and the technology landscape. They work on gaining visibility and support for CRM initiatives by demonstrating how these can address immediate pain points. At higher maturity levels, leaders manage more complex integrations and expansions, ensuring CRM strategies align with business objectives and are flexible enough to adapt to future changes and technological advancements.

Influencing without formal authority in the early stages involves building relationships and advocating for

CRM systems' strategic benefits. Leaders must communicate effectively to overcome resistance and foster early adoption. As the CoE matures, the leader evolves into a strategic adviser who influences stakeholders, senior management, and external partners, steering company-wide strategy through CRM insights.

Strategic alignment in the early stages focuses on translating high-level business objectives into actionable CRM initiatives that demonstrate quick wins. In the mature stage, the CoE leader is deeply involved in shaping company strategies, making sure that CRM capabilities continuously support evolving business models and industry demands. The leader's contributions to strategic alignment are instrumental in the success of the CoE and the company.

Change management initially concentrates on specific projects or departments, establishing a solid foundation for broader CRM adoption. As the CoE grows, change management embeds a continuous improvement and innovation culture, requiring advanced strategies to handle wider-scale transformations.

Accountability and responsibility for CoE leaders start with successfully implementing and adopting CRM projects in specific departments or functions. As CoEs mature, they become responsible for more, expanding to include the sustained performance and evolution of the CRM system across the enterprise, often coordinating with multiple business units and external partners to ensure comprehensive benefits.

Advocating for best practices in the early stages involves establishing standards for data quality, user training, and initial system setups. In the mature stage, the focus shifts to leading industry-wide best practices, pioneering innovations, and setting benchmarks that improve the internal CRM ecosystem and influence broader market practices.

Understanding and planning for seasonal variations in sales and service demand starts with basic analytics and forecasting in the early stages. Mature CoEs use sophisticated predictive analytics and machine learning to dynamically optimize resources and improve customer engagement strategies across various business cycles.

By nurturing skill acquisition and expertise sharing, mature CRM CoEs can effectively navigate these evolving challenges, driving ongoing improvements in customer engagement and contributing to overall organizational success.

Mergers and acquisitions (M&A)

In the dynamic landscape of mergers and acquisitions (M&A), the significance of a well-structured Center of Excellence (CoE) cannot be overstated. However, companies must refrain from involving CoE leaders only late in the integration phase. This oversight delays the critical evaluation of essential business functions such as sales, marketing, and customer service, which are necessary for a successful integration.

A solid customer relationship management (CRM) strategy and a robust Center of Excellence (CoE) framework enable early evaluation of a potential deal. A weak CoE on the seller's side can have significant implications for staffing, potentially raising integration costs and hindering the company's growth trajectory. Conversely, a seller organization that leverages its CRM system's dashboard to present its pipeline, year-to-date sales, and performance metrics to potential buyers can significantly reduce integration risks and enhance future growth prospects.

Involving the CoE early in the M&A process offers several key benefits. First, the CoE can help identify potential synergies between the merging organizations, leading to improved efficiency, cost savings, and increased market share. Second, a well-structured CoE can develop a comprehensive integration plan that addresses various aspects of the merger, including technology, processes, and people. Third, the CoE can assess and mitigate integration risks, such as employee resistance, cultural clashes, and data integration challenges. Additionally, the CoE can lead the change management process, ensuring employees are engaged and aligned with the new organizational structure and culture. Last, the CoE can help organizations realize the total value of the merger by ensuring a successful integration and achieving the desired outcomes.

By involving the CoE early in the M&A process, organizations can significantly increase their chances of successful integration and long-term growth.

Working with external partners and vendors

In large enterprises, engaging with numerous partners, systems integrators, and vendors is a common practice to obtain support for CRM initiatives, spanning areas like strategy, implementation, and development. The Center of Excellence (CoE) plays a critical role from the initial selection phase to successful execution. It is essential that executive leadership, given their deep institutional knowledge and responsibilities in administration, security, and compliance, collaborate closely with CoE leadership to make sure that relevant information is shared with partners, thereby contributing to their success.

It's important to note that the best practices and standards implemented should reflect those of the CoE, rather than consultants assigned to the initiatives. To ensure this, CoE leaders must make consulting companies aware of their best practices and standards. These should be enforced through contracts and during implementation.

The success or failure of partners directly reflects on the company's performance, underscoring the importance of fostering collaborative relationships and shared goals.

The changing direction of CRM vendors

The shift towards industry solutions by organizations like Salesforce and Microsoft Dynamics reflects a growing recognition that businesses need more than just marketing, sales, and service functions to succeed. Companies require comprehensive solutions tailored to their specific industries.

For example, Salesforce's Health Cloud platform offers a range of capabilities specific to health care that go beyond primary sales and service functions. These capabilities include care management, utilization management, provider relationship management, and member management. By leveraging Salesforce's Health Cloud, health care payer organizations can streamline operations, improve patient care, and gain a competitive edge.

This industry-specific approach has several advantages. First, it allows companies to improve collaboration by providing a unified platform that integrates various aspects of their business. This can help break down silos, reduce redundant work, and improve efficiency.

Second, the industry solution approach can help reduce software licensing costs. By investing in a single, comprehensive platform, companies can avoid purchasing multiple-point solutions for different areas of a business, large or small.

Third, the industry-solution approach provides a 360-degree view of the customer status in one system. This can help companies better understand their customers' needs and deliver more personalized experiences.

While the shift towards industry solutions offers significant benefits, it also challenges companies. One challenge is the need to redesign or revamp their business leadership structure. A unified industry solution approach may require companies to consolidate siloed teams and restructure their organizational chart.

Another challenge is adapting to new technologies and processes. Companies that have continued to use traditional, on-premises systems for certain capabilities may need to transition to cloud-based solutions to take advantage of industry-specific CRM features and functionality.

Despite these challenges, the industry solution approach is increasingly becoming the preferred choice for businesses seeking a competitive edge. By partnering with technology-platform providers, companies can access innovative solutions tailored to their specific industry's needs to help them more readily achieve their business goals. This shift towards industry solutions is a significant trend reshaping the CRM landscape, and the companies that embrace this trend will be well-positioned to succeed in the digital age.

Building a CRM CoE movement

Building a CRM CoE internal movement begins with rallying early adopters and gaining endorsements through visible project successes. As the CoE matures, this movement evolves into embedding CRM principles into the corporate culture. This involves ongoing engagement through continuous education, annual CRM collaboration events, community building, and recognition programs.

Connecting with an external ecosystem also follows a developmental trajectory. Early engagement typically focuses on learning and benchmarking against best practices. As the CoE matures, its leaders participate in and lead discussions in external ecosystems, contributing to thought leadership, collaborating on cross-industry initiatives, and influencing CRM technology developments.

As CoEs mature, their leaders must evolve from tactical project managers to strategic visionaries who can navigate complex landscapes, influence without authority, manage far-reaching change, and drive the organization toward sustained CRM excellence. This evolution reflects broader shifts in responsibility, influence, and the scope of impact, both within and beyond the organization.

Conclusion

As CRM Centers of Excellence (CoEs) evolve, so do their leaders' roles—from influencing without authority in the early stages to becoming strategic advisers driving enterprise-wide CRM success. Initially, CoEs focus on quick wins and adoption within departments. Over time, they manage complex integrations, align CRM strategies with broader business goals, and adapt to technological advancements.

Mature CoEs champion industry best practices, fostering continuous innovation and improvement. They optimize resources through predictive analytics, streamline operations, and contribute to top-line growth. Effective change management becomes central, embedding a culture of ongoing adaptation and success.

Leaders in mature CoEs become strategic visionaries, guiding company-wide initiatives, connecting with external ecosystems, and influencing CRM advancements across industries. Ultimately, the CoE brings a unified approach to CRM strategies, ensuring all departments align with business objectives, fostering long-term success and customer-centric innovation.

Conclusion

Creating Meaningful Results

by Velu Palani

Establishing and implementing an industry-standard framework is now a tangible reality. It is within reach. This guide has laid the foundation for understanding, creating, and evaluating CRM CoEs, emphasizing their critical role in optimizing business outcomes through CoE excellence.

This resource is a trusted companion for CRM CoE practitioners and leaders, fostering a community dedicated to continual improvement and advancing crucial discussions in this field. Companies that formalize their CoEs reap substantial returns on their CRM investments. Structured approaches empower these organizations to define clear Key Performance Indicators (KPIs), progressively enhance their CRM systems, and deliver substantial business value.

Drawing from my decade-long experience as a CoE leader, I have seen more companies embrace formal CoEs. However, a persistent challenge remains in balancing delivery and governance. A CoE should enable compliance, security, architecture, platform operations, and release management, as well as provide internal advisory services. It should not be directly responsible for delivery, but it must have the authority to intervene in cases of process or policy breaches. The CoE's influence on such areas as conducting audits, issuing scorecards, and periodically disclosing ROI metrics can only go so far.

Successful outcomes hinge on the CoE's accountability for business results derived from the CRM platform, supported by an executive sponsor. This accountability requires a clear vision, sound strategy, capable leadership, and an experienced team. A CoE's journey toward maturity starts with foundational steps, advances through operational stages, and ultimately arrives at customer centricity, as detailed in our exploration of the CoE structure.

Regular external evaluations of a CoE are crucial for gauging success, tracking progress, and identifying improvement areas. They demonstrate the remarkable capacity of a well-functioning CoE to accelerate ROI realization and transform a company beyond initial expectations.

Whether you are a practitioner, leader, or strategist, immerse yourself in the ongoing dialogue on CRM CoE maturity. Use this comprehensive framework and its practical insights to develop and enhance CRM

Centers of Excellence, paving the way for significant and enduring business transformation. The future of CRM CoEs is bright, and every actor's role in shaping it is pivotal.

Thank you for joining us on this transformative journey.

(For further discussion and auxiliary materials, go to crmcoe.com/book-bonuses.)

Glossary

Agile framework function: Developing and adopting the Agile framework by the organization.

Artificial Intelligence (AI): AI refers to the simulation of human intelligence processes by machines, especially computer systems. These processes include learning, reasoning, and self-correction. AI applications include expert systems, natural language processing, speech recognition, and machine vision.

Business continuity function: Building a resilient CRM infrastructure for business continuity.

Business outcomes: Tangible results and impacts achieved through CRM initiatives within a CoE. These include reducing risk, improving user adoption, lowering operational expenses, increasing the Net Promoter Score, and growing revenue.

Business value: Created by effectively managing customer relationships, driving revenue growth, reducing costs, and enhancing the overall customer experience, contributing to the organization's long-term success.

Capabilities: Specific areas of expertise within each domain that deliver organizational value, including vision and strategy, go-to-market, innovation, PMO (project management office), change management, alignment, risk management, DevOps, and data architecture and governance.

Change control function: Introducing CRM changes systematically and safely.

Continuous development function: Ensuring frequent delivery of well-tested incremental changes to CRM production.

Cost avoidance: Financial strategies that prevent unnecessary expenses by implementing measures that preempt potential costs.

Cost optimization: Analyzing and adjusting spending to maximize value and efficiency, ensuring that each dollar spent contributes effectively to business goals.

Cost reduction: Lowering current expenses and operational costs to improve the organization's financial performance through streamlined processes, eliminating waste, negotiating better terms, and implementing efficient technologies or practices.

CRM CoE (customer relationship management Center of Excellence): A CRM CoE is the team assigned to systematically drive business outcomes via the CRM platform(s). It is the strategic and operational hub responsible for driving CRM excellence, aligning CRM initiatives with business objectives, and fostering continuous improvement in customer relationship management practices.

CRM CoE Framework: A structured approach designed to optimize and manage CRM initiatives, integrating CoEs with overarching business objectives and providing a unified language for assessment and enhancement.

CRM CoE realm of interest: Encompasses the comprehensive scope of the CRM CoE's interests, responsibilities, and areas of impact and control, including the domains of customer centricity, operations, and foundation.

Customer centricity domain: Focuses on aligning CRM initiatives with customer needs and expectations, enhancing customer experiences and satisfaction.

Data architecture and governance capability: Ensures strategic data management, design, and optimization.

Data Cultural Aspect: Emphasizes data quality, integrity, and security to enhance CRM effectiveness.

Data integration function: Efficiently integrates CRM systems with other systems to reduce data silos and enhance customer data analysis.

Data quality function: Prevents and reduces duplicative data, maintaining data cleanliness and usefulness.

Data standardization function: Maintains standardized data naming, definitions, and usage across systems.

DevOps capability: Integrates software development with IT operations to enhance CRM system performance.

Digital revolution: A transformative era that reshaped how organizations operate and interact with customers, driven by advancements in personal computers, the internet, mobile devices, and AI.

Digital transformation: The process by which organizations integrate digital technologies into all business areas, changing how they operate and deliver value to customers.

Domains: Broad categories within the CRM CoE realm of interest that group related capabilities and functions. Domains include foundation, operations, and customer centricity.

Economic optimization: The strategic process of efficiently using resources to achieve the best possible financial outcomes, aligning CRM activities and investments with business objectives to maximize revenue, minimize costs, and enhance overall value.

Experimentation function: Developing a culture of testing new ideas through experimentation.

Five business outcomes: Key benchmarks measuring the effectiveness of CRM strategies: reducing risk, improving user adoption, lowering operational expenses, increasing NPS, and growing revenue.

Five Cultural Aspects: Sets of functions across capabilities and domains shaping the CRM CoE's identity and operational framework, ensuring balanced, sustainable CRM maturity. They include data, people, process, technology, and agility.

Five-Step Process to CRM CoE refinement: A systematic approach to maturing a CRM CoE that involves selecting functions for refinement, drilling down into selected functions, defining ideals, executing improvements, and validating results.

Functions: Actionable task areas within each capability driving the CRM CoE toward strategic goals, including goal setting and prioritizing, transparency, value driven, marketing, sales operation, ideation, experimentation, value creation, talent management, budget, product road map, change control, adoption, Agile framework, workflow/process map, simplification, maintainability, risk mitigation, access control, business continuity, continuous development, testing and automation, talent development, integration, data quality, and data standardization.

Generative AI: A subset of artificial intelligence focused on creating new content or data resembling human-generated output, enhancing creativity, productivity, and innovation in various fields.

Goal setting and prioritizing function: Defining and communicating CRM objectives to ensure alignment with business value.

Hub-and-spoke governance model: A CRM operating model where a central CoE (hub) sets strategy and standards while decentralized teams (spokes) adapt these to their specific needs.

Innovation capability: Evaluating and implementing new methods or products to keep the organization ahead.

Integration function: Managing the effective integration of the CRM with other systems.

Leadership evolution: The shift from traditional, tactical management roles to strategic visionary leadership within CRM CoEs is necessary to navigate complex technological landscapes and drive sustained CRM excellence.

Lowering operational expenses business outcome: Reducing costs associated with CRM operations through efficient processes and resource optimization.

Maintainability function: Maintaining evolving systems without adding complexity or risk.

Marketing function: Focusing on outreach as supported by the CRM.

Nurture skill acquisition and expertise sharing pillar: Developing and managing the CRM CoE team's skills and knowledge to ensure effective CRM system operation and continuous improvement.

Operations domain: Manages and optimizes the operational aspects of CRM implementations, including project management and change management.

People cultural aspect: Fostering a culture of continuous improvement and innovation, focusing on the development and engagement of personnel.

PMO (project management office) capability: Ensures efficient resource allocation and timely project delivery.

Process cultural aspect: The systematic approach to managing CRM workflows and operations, enhancing efficiency, and aligning with strategic goals.

Product road map function: Creating and managing a timeline for delivering business functionalities.

Reducing risk business outcome: Implementing strategies to minimize potential risks associated with the CRM system, ensuring stability and security.

Risk management capability: Identifying and mitigating risks to maintain the integrity of operations.

Risk mitigation function: Limiting or removing risk to or from CRM systems.

ROI realization: The process of achieving and demonstrating the financial benefits of CRM initiatives, supporting continued investment in the CRM system.

Sales operation function: Managing sales processes as supported by CRM.

Service center function: Managing service processes supported by CRM.

Simplification function: Reducing system complexities.

Six Pillars: Core work of the CoE. These include 1) drive strategic alignment 2) establish governance 3) measure ROI realization 4) incorporate collaboration and interoperability 5) hone strategy stewardship and serve as an innovation hub, and 6) nurture acquisition of skills and expertise sharing.

Strategy stewardship and innovation hub pillar: Focused on maintaining strategic consistency and fostering innovation within the CRM CoE.

Talent development function: Creating and maturing employee talent development programs aligned with company needs.

Talent management function: Allocating and managing inside and outside talent to tasks.

Technology cultural aspect: Tools and platforms used in the CRM system emphasizing strategic deployment and alignment with business objectives.

Testing & automation function: Automated/manual CRM testing and other automation of development processes.

Transparency function: Ensuring clear communication and openness across the CRM Center of Excellence realm of interest.

Value creation function: Developing new products, services, or approaches for internal and external business value.

Value driven function: Align CRM strategies with creating tangible business value.

Vision and strategy capability: Developing and articulating clear goals and strategic plans for CRM initiatives.

Workflow/process map function: Articulating, analyzing, documenting, and improving business processes and IT workflows.

References

Preface by Velu Palani

Tom Wong, Liz Kao, and Matt Kaufman, *Salesforce for Dummies*, 6th ed. (Hoboken, NJ: Wiley, 2019).

Preface by Charlie Havens

George Westerman, Didier Bonnet, and Andrew McAfee, *Leading Digital: Turning Technology into Business Transformation* (Boston: Harvard Business Review Press, 2014).

Mark W. Johnson, *Reinvent Your Business Model: How to Seize the White Space for Transformative Growth*, (Boston: Harvard Business Review Press, 2018).

Paul Greenberg, *CRM at the Speed of Light: Social CRM Strategies, Tools, and Techniques for Engaging Your Customers*, 4th ed. (New York: McGraw-Hill, 2010).

Chapter 1: The CRM CoE Framework

AXELOS, ITIL Foundation: ITIL 4 Edition.

Project Management Institute, A Guide to the Project Management Body of Knowledge (PMBOK Guide), 6th ed. (Newtown Square, PA: Project Management Institute, 2017).

Chapter 2: The Cookbook

Vision and Strategy Capability

Darrell K. Rigby, Sarah Elk, and Steve Berez, *Doing Agile Right: Transformation Without Chaos* (Boston: Harvard Business Review Press, 2020).

Eric Ries, *The Lean Startup: How Today's Entrepreneurs Use Continuous Innovation to Create Radically Successful Businesses* (New York: Crown Business, 2011).

Jeff Sutherland, Scrum: *The Art of Doing Twice the Work in Half the Time* (New York: Crown Business, 2014).

Kim Scott, *Radical Candor: Be a Kick-Ass Boss Without Losing Your Humanity* (New York: St. Martin's Press, 2017).

Go-To-Market Capability

John A. Goodman, *Customer Experience 3.0: High-Profit Strategies in the Age of Techno Service* (New York: AMACOM, 2014).

Neil Rackham, *SPIN Selling* (New York: McGraw-Hill, 1988).

Philip Kotler and Kevin Lane Keller, *Marketing Management*, 15th ed. (Boston: Pearson, 2016).

W. Chan Kim and Renée Mauborgne, *Blue Ocean Strategy: How to Create Uncontested Market Space and Make the Competition Irrelevant*, expanded ed. (Boston: Harvard Business Review Press, 2015).

Innovation Capability

Alexander Osterwalder and Yves Pigneur, *Value Proposition Design: How to Create Products and Services Customers Want* (Hoboken, NJ: Wiley, 2014).

Clayton M. Christensen and Michael E. Raynor, *The Innovator's Solution: Creating and Sustaining Successful Growth* (Boston: Harvard Business Review Press, 2003).

Jeff Gothelf and Josh Seiden, *Lean UX: Applying Lean Principles to Improve User Experience* (Sebastopol, CA: O'Reilly Media, 2013).

Nir Eyal, *Hooked: How to Build Habit-Forming Products* (New York: Portfolio, 2014).

Safi Bahcall, *Loonshots: How to Nurture the Crazy Ideas That Win Wars, Cure Diseases, and Transform Industries* (New York: St. Martin's Press, 2019).

Ryan Babineaux and John Krumboltz, *Fail Fast, Fail Often: How Losing Can Help You Win* (New York: TarcherPerigee, 2013)

PMO Capability

Antonio Nieto-Rodriguez, *The Project Revolution: How to Succeed in a Project Driven World* (London: LID Publishing, 2019).

C. Todd Lombardo, Bruce McCarthy, Evan Ryan, and Michael Connors, *Product Road Maps Relaunched: How to Set Direction While Embracing Uncertainty* (Sebastopol, CA: O'Reilly Media, 2017).

Glen S. Gooding, *The IT Financial Management Lifecycle: Budgeting, Costing, Chargeback, and Benchmarking* (London: Kogan Page, 2010).

Greg Horine, *Project Management Absolute Beginner's Guide*, 4th ed. (Indianapolis: Que Publishing, 2017).

Marcus Buckingham and Curt Coffman, *First, Break All the Rules: What the World's Greatest Managers Do Differently* (New York: Simon & Schuster, 1999).

Change Management Capability

John P. Kotter, *Leading Change* (Boston: Harvard Business Review Press, 1996).

Michael E. D. Koenig and David R. Koehler, *The Art of Training Delivery: How to Train Employees Effectively* (New York: Wiley, 2002).

Mike Cohn, *Agile Estimating and Planning* (Upper Saddle River, NJ: Prentice Hall, 2006).

Alignment Capability

Alan Siegel and Irene Etzkorn, *Simple: Conquering the Crisis of Complexity* (New York: Hachette Books, 2013).

Al Decker and Donna Galer, *Enterprise Risk Management: Straight to the Point* (New York: Business Expert Press, 2013).

Gene Kim, Jez Humble, Patrick Debois, and John Willis, *The DevOps Handbook: How to Create World-Class Agility, Reliability, & Security in Technology Organizations* (Portland, OR: IT Revolution Press, 2016).

Gene Kim, Kevin Behr, and George Spafford, *The Phoenix Project: A Novel About IT, DevOps, and Helping Your Business Win* (Portland, OR: IT Revolution Press, 2013).

John Jeston and Johan Nelis, *Business Process Management: Practical Guidelines to Successful Implementations*, 3rd ed. (London: Routledge, 2014).

Philippe Kruchten, Robert Nord, and Ipek Ozkaya, *Managing Technical Debt: Reducing Friction in Software Development* (Boston: Addison-Wesley Professional, 2019).

Quentin Brook, *Lean Six Sigma and Minitab: The Complete Toolbox Guide for Business Improvement*, 5th ed. (Winchester, UK: OPEX Resources, 2014).

Tristan Boutros and Tim Purdie, *The Process Improvement Handbook: A Blueprint for Managing Change and Increasing Organizational Performance* (New York: McGraw-Hill, 2013).

Peter Weill and Jeanne W. Ross, IT Governance: *How Top Performers Manage IT Decision Rights for Superior Results* (Boston: Harvard Business Review Press, 2004).

Risk Management Capability

David Hillson, *The Risk Management Handbook: A Practical Guide to Managing the Multiple Dimensions of Risk* (London: Kogan Page, 2016).

James Lam, *Enterprise Risk Management: From Incentives to Controls*, 2nd ed. (Hoboken, NJ: Wiley, 2014).

Kurt J. Engemann, *The Routledge Companion to Risk, Crisis, and Security in Business* (London: Routledge, 2018).

Leighton Johnson, *Security Controls Evaluation, Testing, and Assessment Handbook* (Waltham, MA: Butterworth-Heinemann, 2015).

Susan Snedaker, *Business Continuity and Disaster Recovery Planning for IT Professionals*, 2nd ed. (Waltham, MA: Syngress, 2013).

DevOps Capability

Jez Humble and David Farley, *Continuous Delivery: Reliable Software Releases through Build, Test, and Deployment Automation* (Boston: Addison-Wesley Professional, 2010).

Red Hat. *'What is CI/CD?'*, Published December 23, 2023. https://www.redhat.com/en/topics/devops/what-is-ci-cd#:~:text=CI%2FCD%2C%20which%20stands%20for,a%20shared%20source%20code%20repository.

Lisa Crispin and Janet Gregory, *Agile Testing: A Practical Guide for Testers and Agile Teams* (Boston: Addison-Wesley Professional, 2009).

Ruth C. Clark, *Developing Technical Training: A Structured Approach for Developing Classroom and Computer-based Instructional Materials* (Silver Spring, MD: International Society for Performance Improvement, 2012).

Data Architecture Capability

Anthony David Giordano, *Data Integration Blueprint and Modeling: Techniques for a Scalable and Sustainable Architecture* (Upper Saddle River, NJ: IBM Press, 2010).

CyberArk. *"The Seven Types of Non-human Identities to Secure."* Last modified May 16, 2023. https://www.cyberark.com/resources/blog/the-seven-types-of-non-human-identities-to-secure.

John W. Foreman, *Data Smart: Using Data Science to Transform Information into Insight* (Hoboken, NJ: Wiley, 2013).

Jack E. Olson, *Data Quality: The Accuracy Dimension* (San Francisco: Morgan Kaufmann, 2003).

Ralph Kimball and Margy Ross, *The Data Warehouse Toolkit: The Definitive Guide to Dimensional Modeling*, 3rd ed. (Indianapolis: Wiley, 2013).

Thomas H. Davenport and Jinho Kim, *Keeping Up with the Quants: Your Guide to Understanding and Using Analytics* (Boston: Harvard Business Review Press, 2013).

Chapter 3: Iterative Transformation

Esther Derby and Diana Larsen, *Agile Retrospectives: Making Good Teams Great* (Raleigh, NC: Pragmatic Bookshelf, 2006).

Jeff Sutherland, *Scrum: The Art of Doing Twice the Work in Half the Time* (New York: Crown Business, 2014).

Chapter 4: Value Realization With CRM CoE

Douglas W. Hubbard, *How to Measure Anything: Finding the Value of Intangibles in Business*, 3rd ed. (Hoboken, NJ: Wiley, 2014).

Gene Kim, *The Unicorn Project: A Novel about Developers, Digital Disruption, and Thriving in the Age of Data* (Portland, OR: IT Revolution Press, 2019).

Thijs Homan, *Organization Dynamics in the Network Era* (London: Springer, 2017).

William D. Pace, *Return on Investment (ROI): Basics for Consultants* (Boca Raton, FL: CRC Press, 2009).

Chapter 5: The Cost of Running the Ecosystem

Glen S. Gooding, *The IT Financial Management Lifecycle: Budgeting, Costing, Chargeback, and Benchmarking* (London: Kogan Page, 2010).

Sanjeev Purushotham, *The Economics of IT Cloud Computing: Billing, Capacity, and Costing* (New York: Springer, 2018).

Todd Tucker, *Technology Business Management: The Four Value Conversations CIOs Must Have with Their Businesses* (Bellevue, WA: TBM Council, 2016).

Nucleus Research, "Salesforce ROI Case Study: Pearson Education," June 15, 2021 - ROI Case Studies

V91, https://nucleusresearch.com/research/single/salesforce-roi-case-study-pearson-education/.

Gartner Research. "Use Total Cost of Ownership to Optimize Costs and Increase Savings." Last modified January 19, 2018. https://www.gartner.com/en/documents/3847267.

Forrester Consulting, The Total Economic Impact™ of GE Digital's Asset Performance Management, commissioned by GE Digital, July 2022, https://www.ge.com/digital/sites/default/files/download_assets/forrester-total-economic-impact-ge-digital-apm.pdf.

IIDC, "ROI from CRM Ranges from 16% to 1000%," The Wise Marketer, accessed November 14, 2024, https://thewisemarketer.com/roi-from-crm-ranges-from-16-to-1000/.

Chapter 6: CoE Leadership

Adam Riccoboni, *AI Age: How Artificial Intelligence is Transforming Organizations* (London: TechPress, 2020).

Brené Brown, *Dare to Lead: Brave Work. Tough Conversations. Whole Hearts.* (New York: Random House, 2018).

Gary Gruver and Tommy Mouser, *Leading the Transformation: Applying Agile and DevOps Principles at Scale* (Portland, OR: IT Revolution Press, 2015).

Jeanne Liedtka, *Design Thinking for the Greater Good: Innovation in the Social Sector* (New York: Columbia University Press, 2017).

Mark W. Johnson, *Reinvent Your Business Model: How to Seize the White Space for Transformative Growth* (Boston: Harvard Business Review Press, 2018).

Patrick Lencioni, *The Five Dysfunctions of a Team: A Leadership Fable* (San Francisco: Jossey-Bass, 2002).

Satya Nadella, *Hit Refresh: The Quest to Rediscover Microsoft's Soul and Imagine a Better Future for Everyone* (New York: Harper Business, 2017).

Simon Sinek, *Leaders Eat Last: Why Some Teams Pull Together and Others Don't* (New York: Portfolio, 2014).

Sunil Gupta, *Driving Digital Strategy: A Guide to Reimagining Your Business* (Boston: Harvard Business Review Press, 2018).

Thomas M. Siebel, *Digital Transformation: Survive and Thrive in an Era of Mass Extinction* (New York: RosettaBooks, 2019).

Chapter 7: CRM CoE Maturity

Peter M. Senge, *The Fifth Discipline: The Art & Practice of The Learning Organization* (New York: Crown Business, 2006).

Verne Harnish, *Scaling Up: How a Few Companies Make It...and Why the Rest Don't* (Ashburn, VA: Gazelles Inc., 2014).

David H. DeWolf and Jessica S. Hall, *The Product Mindset: Succeed in the Digital Economy by Changing the Way Your Organization Thinks* (New York: HarperCollins Leadership, 2019).

Chapter 8: Navigating the Enterprise

Jeanne W. Ross, Peter Weill, and David C. Robertson, *Enterprise Architecture as Strategy: Creating a Foundation for Business Execution* (Boston: Harvard Business Review Press, 2006).

Jürgen Großmann, *Enterprise Architecture Frameworks: The Agile Way to Design Enterprise Architectures* (Berlin: Springer, 2019).

Marc Lankhorst, *Enterprise Architecture at Work: Modeling, Communication, and Analysis*, 4th ed. (Berlin: Springer, 2017).

Stefan Bente, Uwe Bombosch, and Shailendra Langade, *Collaborative Enterprise Architecture: Enriching EA with Lean, Agile, and Enterprise 2.0 practices* (Waltham, MA: Morgan Kaufmann, 2012).

About the Authors

Velu Palani and Charlie Havens have collaborated for over twelve years, bringing their expertise to sizable corporations and smaller nonprofits. Their combined skills in transforming business needs into IT solutions, particularly for Salesforce, have fostered a partnership grounded in innovation and practical application. This book is borne of their shared concern that Centers of Excellence frequently have to reinvent themselves anew within each company without the advantage of common frameworks, standards, and terminology. The recommendations and conclusions in this book mark the culmination of their work together.

Velu Palani, a visionary leader in digital transformation and CRM optimization, brings a wealth of experience. As the Executive Director of Salesforce Platform Services & CoE at Health Care Service Corporation, he has overseen major Salesforce implementations across several states, demonstrating his ability to handle complex projects. With a Salesforce customer history dating from June 2000, Velu has amassed over two decades of experience with Salesforce.com, undertaking projects for a diverse clientele, from small to large corporations. His interest matches his enthusiasm for maximizing CRM investments in integrating artificial intelligence within CRM Centers of Excellence to boost functionality and efficiency. Velu founded CRMCoE .com, a roundtable platform for CRM CoE leaders to exchange best practices. His leadership is not confined to technical roles; he also participates in CRM professional mentoring and enjoys enriching his global perspective through travel.

Charlie Havens, known for his keen ability to understand business needs, is driven by a deep passion for his work. Now, in the later stages of his career, Charlie engages primarily for enjoyment, applying his expertise to enhance CRM Centers of Excellence and optimize operating costs. A CRM governance and execution specialist, Charlie previously helped build the Salesforce Platform Strategic Enablement Team at Health Care Service Corporation. He also established the Chicago Salesforce Partner User Group, fostering a community that thrives on shared knowledge and best practices in Salesforce partnership. Charlie's experience includes managing substantial CRM projects, such as implementing Salesforce Sales Cloud solutions for both B2B and B2C sectors at T. Rowe Price. Beyond his professional pursuits, Charlie delights in mushroom foraging and sharing his passion for the natural world. He also creates a pretty good cassoulet (an elaborate French stew of various meats and beans).

Index

P

R

S

T

U

V

W

www.ingramcontent.com/pod-product-compliance
Lightning Source LLC
Chambersburg PA
CBHW051755200326
41597CB00025B/4562